Environmental Impact of Aviation and Sustainable Solutions

Edited by Ramesh K. Agarwal

Published in London, United Kingdom

IntechOpen

Supporting open minds since 2005

Environmental Impact of Aviation and Sustainable Solutions
http://dx.doi.org/10.5772/intechopen.75289
Edited by Ramesh K. Agarwal

Contributors
Vivekananda Mukhopadhyay, Bo Jin, Michael Ellis, Karl Neidert, Oluwaferanmi Oguntona, Sergej Tezak, Constantin Sandu, Radu-Constantin Sandu, Cristian-Teodor Olariu, Dianodng Ren, Lance Leslie, Kateryna Synylo, Oleksander Zaporozhets, Amir Esmael Forouhid, Haifeng Liu, Hui Zhao, Zhi X. Chen, Ivan Langella, N Swaminathan, Xiaochen Li

First published in London, United Kingdom, 2020 by IntechOpen
IntechOpen is the global imprint of INTECHOPEN LIMITED, registered in England and Wales, registration number: 11086078, 7th floor, 10 Lower Thames Street, London, EC3R 6AF, United Kingdom
Printed in Croatia

British Library Cataloguing-in-Publication Data
A catalogue record for this book is available from the British Library

Additional hard and PDF copies can be obtained from orders@intechopen.com

Environmental Impact of Aviation and Sustainable Solutions
Edited by Ramesh K. Agarwal
p. cm.
Print ISBN 978-1-83962-357-8
Online ISBN 978-1-83962-358-5
eBook (PDF) ISBN 978-1-83962-359-2

For EU product safety concerns:
IN TECH d.o.o., Prolaz Marije Krucifikse Kozulić 3, 51000 Rijeka, Croatia, info@intechopen.com or visit our website at intechopen.com.

Meet the editor

Professor Ramesh K. Agarwal is William Palm Professor of Engineering at Washington University in St. Louis, USA. From 1994 to 2001, he was Sam Bloomfield Distinguished Professor and Executive Director of the National Institute for Aviation Research at Wichita State University in Kansas. From 1978 to 1994, he worked in various scientific and managerial positions at McDonnell Douglas Research Laboratories in St. Louis; he became the Program Director and McDonnell Douglas Fellow in 1990. Dr. Agarwal received a PhD in Aeronautical Sciences from Stanford University in 1975, an MS in Aeronautical Engineering from the University of Minnesota in 1969, and a BS in Mechanical Engineering from the Indian Institute of Technology, Kharagpur, India in 1968. Dr. Agarwal has worked in computational fluid dynamics and its applications to problems in aerospace and mechanical engineering, and energy and the environment.

Contents

Preface

Environmental Impact of Aviation and Sustainable Solutions is a compilation of review and research articles in the broad field of aviation and the environment. It is divided into three sections.

The first section contains four chapters that provide a state-of-the-art review of aircraft structural designs and materials for contributing towards environmentally responsible aviation.

In the second section, five chapters cover topics ranging from combustor modeling, atomization, and airport pollution to mitigation of sonic boom and street noise pollution in the vicinity of the airport.

In the third section, two chapters describe emission mitigation strategies, while the last chapter of the book describes the contributions of Russian aviation pioneer Robert Bartini whose designs could be considered as environmentally desirable.

This book discusses a wide variety of topics related to environmental impact of aviation and some suggestions to reduce the aviation industry's adverse impacts. It serves as a useful reference for both researchers and students interested in learning about various aspects of aviation and the environment as well as potential solutions to reduce negative environmental impacts.

Ramesh K. Agarwal
Washington University in St. Louis,
USA

Section 1

Structures, Materials and Manufacturing

Chapter 1

Structural Analysis of Electric Flight Vehicles for Application of Multifunctional Energy Storage System

Vivek Mukhopadhyay

Abstract

The Multifunctional Structures for High Energy Lightweight Load-bearing Storage (M-SHELLS) research project goals were to develop M-SHELLS, integrate them into the structure, and conduct flight tests onboard a remotely piloted small aircraft. Experimental M-SHELLS energy-storing coupons were fabricated and tested for their electrical and mechanical properties. In this chapter, finite element model development and structural analyses of two small test aircraft candidates are presented. The component weight analysis from the finite element model and test measurements were correlated. Structural analysis results with multifunctional energy storage panels in the fuselage of the test vehicle are presented. The results indicate that the mid-fuselage floor composite panel could provide structural integrity with minimal weight penalty while supplying electrical energy. Structural analyses of the NASA X-57 Maxwell electric aircraft and an advanced aircraft fuselage structure are also presented for potential application of M-SHELLS. Secondary aluminum structure in the fuselage subfloor and cargo area are partially replaced with reinforced five-layer composite panels with M-SHELLS honeycomb core. The fuselage weight reduction associated with each design without risking structural integrity are described. The structural analysis and weight estimation with composite M-SHELLS panels in the fuselage floor indicates 3.2% structural weight reduction, but with increased stress.

Keywords: advanced composite, multifunctional structure, green aviation, electric flight vehicle design, finite element analysis, honeycomb panel, electrical energy storage, structural weight estimation and reduction, aircraft design

1. Introduction

For sustainable green aviation, the innovative electric flight vehicle structures should be lighter, yet safer than the existing technology can offer, in order to reduce the overall weight and subsequently fuel consumption and emission. This chapter describes structural design of advanced electric flight vehicle concepts, which are potential candidates to meet some of the environmental friendly performance goals. Under the NASA Aeronautics Research, Convergent Aeronautical Solution Program, Glenn Research Center (GRC) has been leading Multifunctional Structures for High Energy Lightweight Load-bearing Storage (M-SHELLS) research. The technology of integrating load-carrying structures with electrical energy storage capacity has the potential to reduce the overall weight of future electric aircraft. Langley Research Center (LaRC) along with GRC fabricated and tested lightweight, laminated honeycomb composites with special anode, cathode, and separator materials that are dually capable of generating electrical power and carrying mechanical loads. Storing and releasing electrical energy with hybrid super-capacitors combined with advanced composite structures has the potential to reduce both the charging time and overall weight. Krause and Loyselle [1] at GRC proposed developing, analyzing, and testing this multifunctional structures technology. The Materials and Electro-chemistry Division at GRC has conducted extensive research on multifunctional structural composites that are capable of generating electrical power and carrying mechanical loads.

Figure 1 shows a roadmap of the multifunctional structures technology development and systems analysis [2]. At GRC, advanced multifunctional composite laminate and hybrid super-capacitor energy storage systems are being developed. Numerical models of electrochemical reactions and energy storage concepts are also being developed at GRC. Newman [3] presented the specific energy and specific power characteristics of existing fuel cell and battery technologies and conventional energy sources in the Ragone plot (**Figure 1a**). The initial performance goal for the M-SHELLS system was to demonstrate a specific energy of 75 Wh/kg at a specific power of 1000 W/kg. These modest M-SHELLS specific energy and power targets are also shown in **Figure 1a**. An expanded view of the Ragone plot is shown in **Figure 2** for additional discussion. The honeycomb sandwich structure for the M-SHELLS concept is shown in **Figure 1b**. Specimens were fabricated and tested in the structures concept laboratory at GRC and LaRC to characterize both the electrochemical and mechanical properties. **Figure 1c** shows one tensile test result of an initial single layer experimental M-SHELLS honeycomb specimen.

The remotely piloted small airplane, named *Tempest*, developed by UASUSA Inc., was acquired for retrofitting with a multifunctional system to provide partial power and augment the existing Lithium-Polymer (Li-Po) battery (**Figure 1d**). The Li-Po battery provides 4 amperes of current for peak power during catapult launching and 2 amperes of continuous current for cruise power. A separate battery supplies steady power to the flight control system. The objective of the flight test project was to augment the present 18.5-volt Li-Po battery with an M-SHELLS power pack to demonstrate its functionality and flight worthiness. Although the planned flight test was eventually cancelled due to project constraints, the initial structural model development and associated structural analyses are presented.

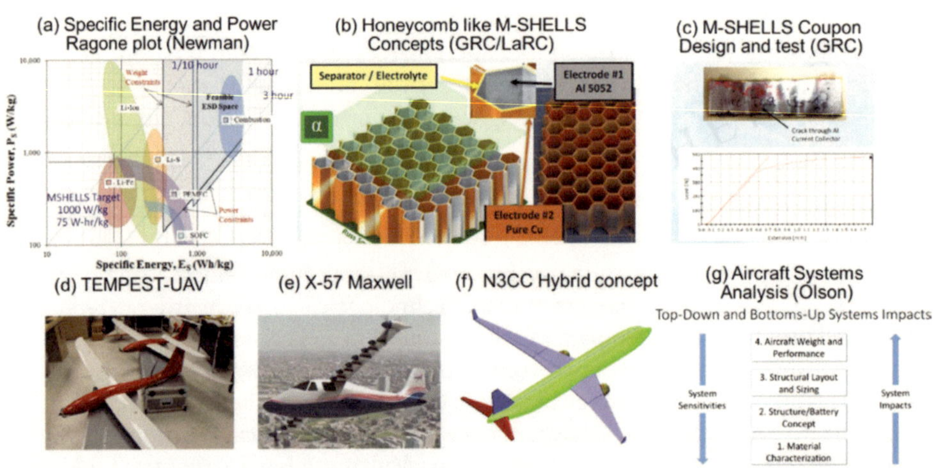

Figure 1.
Multifunctional load bearing structure and systems analysis roadmap.

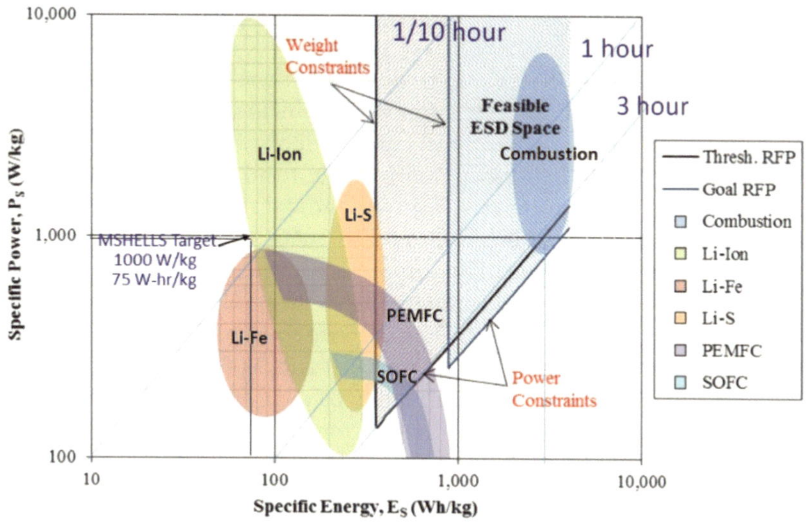

Figure 2.
Ragone plot for specific energy and specific power characteristics of energy source devices.

Figure 1e shows the NASA X-57 Maxwell experimental test aircraft concept [4] with a distributed electric propulsion system that has 12 electric-motor-driven propellers on the high-lift wing. The synchronized motors are powered by a 358 kg battery pack. Presently, construction of the X-57 Maxwell test vehicle is occurring under the Scalable Convergent Electric Propulsion Technology Operational Research (SCEPTOR) project. The X-57 Maxwell vehicle will test the performance of this specially designed wing with distributed electric propulsion to evaluate mission benefits for this class of vehicle. Structural analysis of the fuselage floor modeled with a reinforced M-SHELLS composite panel is briefly described.

As a final application, structural and aircraft systems analysis for the NASA N+3 Technology Conventional Configuration (N3CC) derivative with hybrid-electric propulsion (**Figure 1f**) were conducted by Olson and Ozoroski [2] in order to predict the multifunctional performance and weight benefits of the M-SHELLS technology (**Figure 1g**). In this report, secondary aluminum structure in the N3CC fuselage sub-floor and cargo area are partially replaced with M-SHELLS composite panels for structural stress and weight analysis.

Newman [3] presented an extensive feasibility and design study of a small, manned aircraft with electric powered propulsion. His report included the range of specific energy and specific power characteristics for existing Lithium-based batteries, Proton-Exchange Membrane Fuel Cells (PEMFC), Solid Oxide Fuel Cells (SOFC), and aviation fuel. **Figure 2** is his summary plot of the specific power and energy specifications, which is often referred to as a Ragone plot. Newman concluded that, besides conventional combustion, PEMFC and SOFC were the only two feasible energy source devices given the selected set of mission and aerodynamic (weight and power) constraints and the design specifications for his project. The initial performance goal for the M-SHELLS battery system was to demonstrate a specific energy of 75 Wh/kg at a specific power of 1000 W/kg. These M-SHELLS energy and power targets are superimposed on Newman's plot in **Figure 2**. While this target is modest compared to Li-Ion, Li-Fe, and Li-S based batteries, the main advantage of the M-SHELLS technology is that it could replace part of the load bearing structure, particularly in small drones and in lightly loaded fuselage structure of experimental electric aircraft such as the X-57 Maxwell.

2. M-SHELLS coupon test

The proposed M-SHELLS research goals were to develop test specimens and subcomponents, integrate them into a small test vehicle structure, and conduct low-risk flight tests. The M-SHELLS test coupons in the form of honeycomb panels were fabricated and tested by Russell Smith (LaRC) and Brett Bednarcyk (GRC) for mechanical and electrical properties. **Figure 3** shows the normal compression load shakedown test of a small, stabilized aluminum honeycomb coupon fabricated for mechanical property assessment. The compressive crushing strength and compressive modulus were computed and compared with the published characteristics of a Hexcel 1/4-5052-0.002 honeycomb. The flatwise compression modulus of the aluminum honeycomb coupon with 1/4-inch cell and 0.002-inch foil thickness is 139,000 psi and the crushing strength is 436 psi. The published in-plane shear modulus of the Hexcel 1/4-5052-0.002 honeycomb is 66,000 psi and the shear strength is 300 psi in the length direction. In the width direction, the in-plane shear modulus is 30,000 psi and the shear strength is 120 psi. Since the normal compression strength test result and Hexcel published data were very close, the mechanical properties of Hexcel honeycomb were used by Olson and Ozoroski [2] for the initial structural and multifunctional performance benefit analysis of the N3CC derivative with hybrid-electric propulsion. They also accounted for the additional weight of core material required to complete the energy storage functionality.

Figure 4 shows the in-plane tensile load versus extension plot from an initial tensile test of an early M-SHELLS active coupon prototype with anode/cathode elements and electrolytes. The honeycomb test coupon dimensions were 6.0 inch (150 mm) in length, 2.0 inch (50.8 mm) in width, and 1.0 inch (25.4 mm) in depth. The face-sheets were 0.002 inch thin aluminum foil. The electrical tests were conducted at NASA Glenn Research Center. Considering only the linear part of the deformation, a 90 lb (400 N) load produces an extension of 0.6 mm. Thus, relative to the unloaded specimen, the linear elastic strain was 0.004 at the 90 lb (400 N) load. The specimen yielded beyond the 400 N load and developed a crack at 480 N. The linear Young's modulus (stress/strain) was computed to be 11,188 psi (77.52×106 N/m^2). The corresponding in-plane shear modulus was 4024 psi for the Poisson's ratio of 0.39. The in-plane tensile and shear modulus computed from the coupon test results were very low for flight application. Hence, for the present analysis, additional outer face-sheets were added on each side to add strength to the honeycomb core (**Figure 1b**). Several detailed finite element models (FEM) of three flight vehicles were developed having certain fuselage areas replaced with this reinforced composite panel having a honeycomb core. Structural analyses of these models are described. The complete summary of all material properties used in this

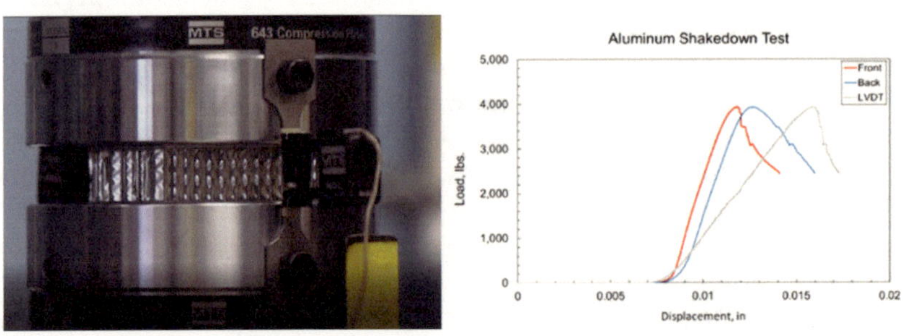

Figure 3.
Normal-compression load shakedown test of a small, stabilized aluminum honeycomb coupon fabricated for mechanical property assessment.

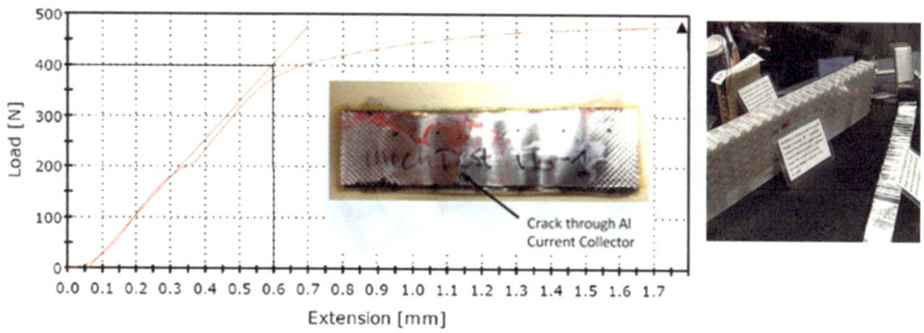

Figure 4.
Initial tensile test result of an experimental M-SHELLS coupon prototype.

chapter are presented in Appendix A. The M-SHELLS panel design properties and computed density are presented in Appendix B.

3. Flight test vehicle structural model development

Initially, several low-cost, small model aircraft were considered for finite element analysis and simulation, with multifunctional lightweight composite panels replacing part of the wing and fuselage structure. A remotely piloted small aircraft was selected with a 127 inch wingspan and a takeoff weight of 16 lb. Adequate details about the internal structure and fabrication of this model airplane were not known, so a notional FEM of this small aircraft was quickly developed for initial structural analysis with design flight loads. **Figure 5** shows a preliminary structural model development of a similarly sized small hobby model airplane, which offered an initial low-risk candidate for flight testing of the M-SHELLS specimen. A typical wing FEM with a standard two-spar and rib configuration was initially developed. This structural arrangement would enable easy integration of small test coupons, between the two spars in the inboard section, close to the electric motor in the fuselage nose. The test specimen could also be integrated into the fuselage floor.

Figure 6 shows the wing deflection and strain distribution from initial structural analysis of the wing in level flight. The analysis assumed front and rear spar thicknesses of 0.15 inch with advanced composite material properties [5]. The linear elastic property values used for the front and rear spar are as follows: Young's modulus 9,750,000 psi, shear modulus 2,570,000 psi, and mass density 0.06 lb/in^3. The wing, fuselage, horizontal tail, and vertical tail skin thicknesses were 0.04 inch and were made of standard thermoplastic material. The linear elastic properties are as follows: Young's modulus 290,075 psi, shear modulus 47,250 psi, and mass density 0.04 lb/in^3. The wing deflections and skin strain distributions shown are with a fixed wing root and a 16 lb lift load, distributed elliptically along the wing. The maximum deflection and nodal strain were 1.95 inches at the wing tip and 0.00106 at the wing root, respectively. With this two-spar wing construction, the maximum wing-tip deflection and strain values at level cruise flight were considered high for a model airplane. The two-spar wing FEM weight was calculated to be 4.63 lb. The fuselage weight, with empennage, was calculated to be 3.8 lb.

When NASA Langley acquired two UASUSA-manufactured remotely piloted aircraft named "Tempest" for the planned flight test, additional information on the internal construction of the physical model was available. A *Tempest* model was dismantled to observe the internal construction at the wing root. The weight of each component of the disassembled model was also measured. Since the material properties of the *Tempest* wing and other model parts were not known, a bench test

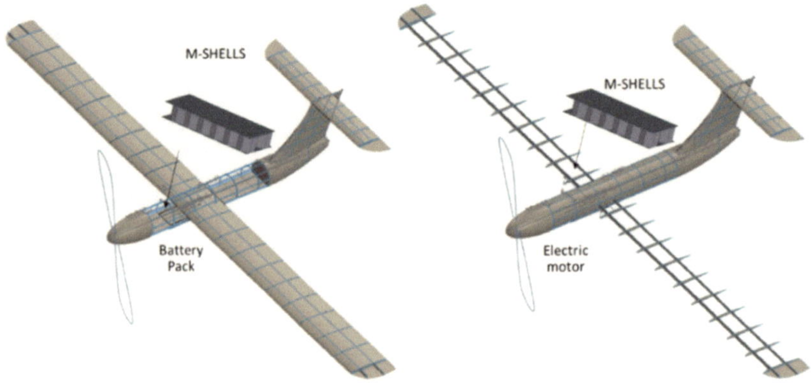

Figure 5.
Preliminary structural model development of the two-spar wing airplane.

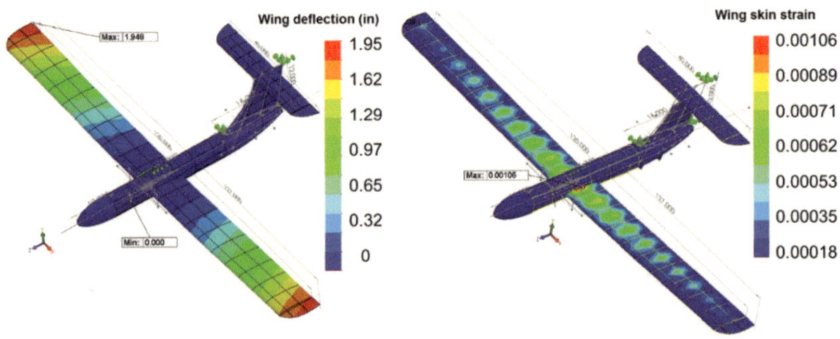

Figure 6.
Wing deflection and strain of the two-spar wing model airplane.

was performed to evaluate the wing deflection and stiffness under a simulated lift load. Gregory Howland and David Hare performed the bench load-deflection test at the NASA Langley model shop on a layout table. The loading configuration was based on the test setup scheme shown in **Figure 7**. The model was inverted and then leveled and supported by two foam blocks. The wing load application points were positioned at 24 inches from the centerline. Eight-pound weights were placed on the right and left wings symmetrically at those reference points. The average wing-tip displacement was ~0.94 of an inch. The load was removed from each wing and then the loading was repeated. The second time, the average wing-tip deflection was 0.96 of an inch. The inset photos in **Figure 7** show the bench test arrangement in the NASA Langley model shop.

Upon close examination of the model with the canopy removed, it was observed that the *Tempest* wing is constructed as two symmetric pieces of hollow, molded composite that are joined together with a short central stub-spar and two solid root-rib pieces, each 2 inches wide. **Figure 8** shows the *Tempest* wing construction. A new finite element model of the wing was developed to represent this construction. The central stub-spar and two wide ribs were modeled with solid advanced composite material properties as before. The molded fiberglass skin of the two wings was modeled as 0.025 inch thin composite material. The rest of the model used custom thermoplastic material.

The horizontal tail skin and ribs were modeled as 0.02 inch thin molded thermoplastic. The fuselage and vertical tail skins and ribs were modeled with 0.04 inch thin thermoplastic. The horizontal and vertical tail twin-spar thicknesses were

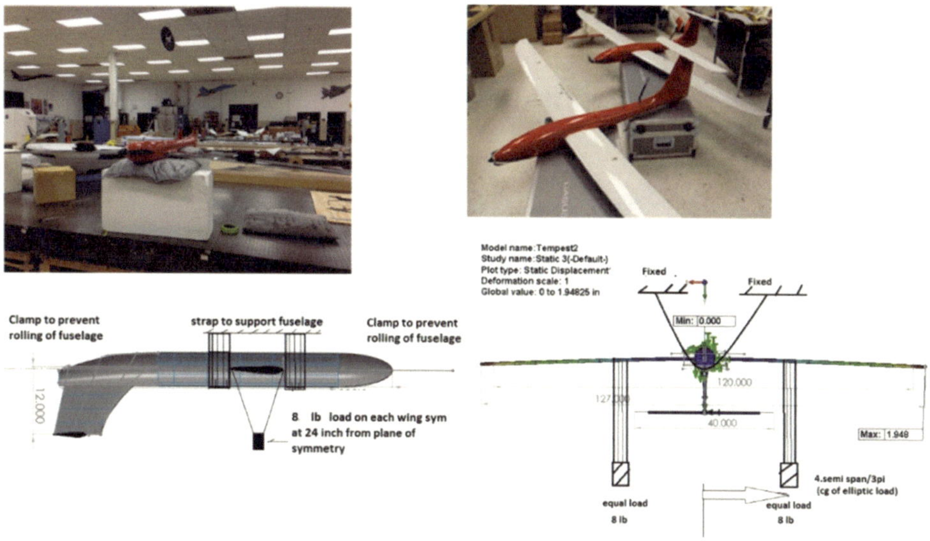

Figure 7.
Wing deflection test of the tempest aircraft with 16 lb total lift load on the wing.

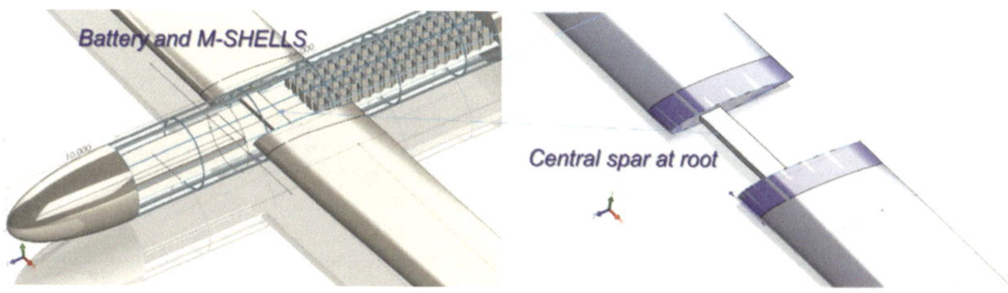

Figure 8.
Structural model and wing root internal detail of the tempest aircraft.

0.10 inch and 0.15 inch, respectively. **Figure 9** shows the wing deflection and nodal strain distributions from the FEM analysis with level flight load, assuming a 16 lb takeoff gross weight. See Appendix A for all the material elastic properties and density used in this chapter. With the improved FEM of the wing structure, the wing-tip deflection was 1.11 inch and the maximum strain at the wing root was 0.00067. The strain values were noted to be well within the allowable limits. The wing-tip deflection was closer to the experimental results than the preliminary FEM analysis results with the two-spar wing (**Figure 6**). This improved FEM analysis result was considered satisfactory for the structural component weight estimation.

Table 1 shows the measured component weights of the test vehicle and estimated weight for the initial two-spar wing model and the improved model of the **Tempest** wing. Some of the structural component weights and the electronic system weight inside the fuselage could not be measured separately, since the fuselage and vertical tails are molded as a single part. Hence, the weights of those components are grouped together in **Table 1**. The two-spar wing weight was estimated to be 4.63 lb. With the better FEM of **Tempest**, the estimated total wing weight of 3.54 lb is closer to the measured combined weight of 3.46 lb for its right and left wings and stub spar. The measured fuselage weight, 5.62 lb, included the co-molded vertical tail and electronic components inside the fuselage. It compared well with the improved FEM combined weight of the fuselage and vertical tail, including an estimated 2 lb weight for electronic components, telemetry system, and motors.

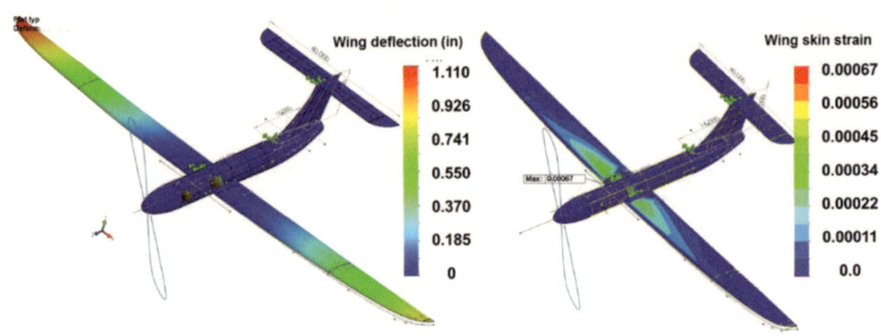

Figure 9.
Wing deflection and strain of the improved finite-element model of the test vehicle in level flight.

Component	Measured Weight (lbs)		FEM model with two spar wing (lb)		Improved Tempest FEM (lb)		Notes
Electronic			2.00	lb	2.00	lb	Estimated
Fuselage + vtail	5.62	lb	3.01	lb	3.01	lb	fus+vt w/o electronic
Canopy	0.27	lb	0.18	lb	0.18	lb	Estimated
Horizontal Tail	0.43	lb	0.66	lb	0.66	lb	
Wing	3.14	lb	4.63	lb	3.54	lb	with stub spar
Stub Spar	0.32	lb					
Flight Battery	2.30	lb	2.30	lb	2.30	lb	from 1st col.
Ballast Required for C.G	1.23	lb	1.23	lb	1.23	lb	from 1st col.
Baseline Op. Wt Total	13.30	lb	14.00	lb	12.91	lb	lb with electronic

Table 1.
Comparison of component weights of the tempest test vehicle, initial two-spar wing model, and improved tempest FEM.

The performance goal for the M-SHELLS development was to demonstrate a specific power of 1000 W/kg at an energy density of 75 Wh/kg. The flight test goal was to augment the existing Li-Po battery with 33% of the required energy for 30 minutes of flight or, equivalently, to supply the full electrical energy for 10 minutes of level flight. The Li-Po battery capacity is 7600 mAh and it provides 7.4 volts with two 3.7 volt cells in series. With a gross weight of 2.3 lb (1.04 kg), the energy density of the Li-Po battery is 55 Wh/kg. The ideal power required by the aircraft at cruise is computed from weight × velocity/(L/D), where L/D is the lift-to-drag ratio. Considering the propeller and motor efficiencies, the total power required to be supplied to the electric motor spinning the propeller is:

$$\text{Power Required} = \text{weight} \times \text{velocity}/\left[L/D \times (\text{propeller efficiency}) \times (\text{motor efficiency})\right] \quad (1)$$

For the *Tempest* test vehicle, let us assume a baseline cruise weight of 20 lb (88 N), a cruise velocity of 40 mph (17.9 m/s), and a typical L/D of 20. Assuming a motor efficiency of 85% and a propeller efficiency of 80%, the power required = 88 × 17.9/(20 × 0.85 × 0.80) = 116 W and the energy required for 10 minutes of level flight is (116 × 10/60) = 20 Wh. Hence, ideally, 0.58 lb (20/75 kg) of M-SHELLS material could provide full power for 10 minutes of level flight. The actual weight of the M-SHELLS power package would depend on the flight test voltages and current demand of the electric motor and the ability to package each unit in suitable series and parallel configurations to match the available power supply and required power demand.

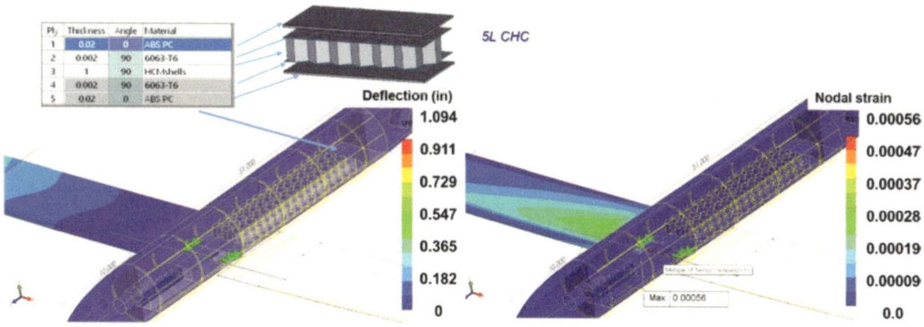

Figure 10.
Tempest FEM analysis with M-SHELLS composite panel fuselage floor.

The structural deflection and nodal strain distribution from the FEM analysis results of the *Tempest* vehicle with a lightweight M-SHELLS composite panel replacing the fuselage floor are shown in **Figure 10**. The five-layer bonded sandwich panel consisted of 0.02 inch thermoplastic sheet for insulation on the outer faces, 0.002 inch aluminum sheet on the inner faces and 1.0 inch deep honeycomb M-SHELLS core. The original fuselage floor weight was 0.32 lb. One stack of this five-layer sandwich energy storage panel replacing 180 in^2 of mid-fuselage floor would weigh 1.25 lb. The mid-fuselage floor composite, multifunctional panel would provide both structural integrity and supply electrical energy to supplement the existing Li-Po battery of this vehicle.

4. NASA X-57 Maxwell test vehicle

Under the Scalable Convergent Electric Propulsion Technology Operational Research (SCEPTOR) project, the X-57 Maxwell test vehicle wing is presently being constructed at NASA Armstrong Flight Research Center. **Figure 1e** showed the NASA X-57 Maxwell experimental test aircraft concept [4] with a distributed electric propulsion system featuring 12 electric-motor-driven propellers on an innovative high-lift wing. The X-57 Maxwell vehicle will test the performance of this specially designed wing with distributed electric propulsion in order to evaluate mission benefits for this class of vehicle.

Figure 11 shows the weight breakdown of the NASA X-57 Maxwell experimental test aircraft. The original wing of the Italian *Tecnam P2006T* aircraft will be replaced with a specially designed distributed electric propulsion wing with 12 electric-motor-driven propellers. The wing-tip propellers help reduce the induced drag from the tip vortex. The synchronized motors are powered by a 358 kg Nickel-Cobalt-Aluminum (NCA) battery pack. The electric power system is organized into eight battery modules, split into two packs with 4 battery modules and a control module each. Cooling is provided through 18,650 cells spaced evenly, 4 mm apart. The NCA cells provide sufficient energy density and the required discharge rate for the flight test mission. Each pack supplies 47 kWh of useful energy, with a peak discharge power of 132 kW. The total battery package weight is estimated to be 790 lb (358 kg), or 26% of the total aircraft takeoff gross weight of 3006 lb (1364 kg). The aluminum fuselage weight is 302 lb (136 kg), and the total estimated structure weight without the landing gear is 738 lb (335 kg).

Figure 12 shows initial power requirement estimates for the standard mission of the X-57 Maxwell [6] flight test vehicle. The energy requirement for each phase of the mission is obtained by integrating the power requirement over time

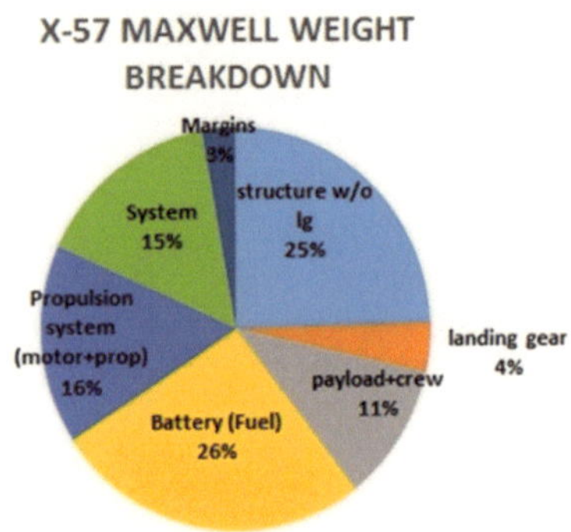

Category	Weight %	weight (lb)	weight (kg)
structure w/o lg	25%	738 lb	335 kg
landing gear	4%	119 lb	54 kg
payload+crew	11%	333 lb	151 kg
Battery (Fuel)	26%	790 lb	358 kg
Propulsion system (motor+prop)	16%	480 lb	218 kg
System	15%	460 lb	209 kg
Margins	3%	86 lb	39 kg
Total	100%	3006 lb	1364 kg

Figure 11.
Component weight fractions for the X-57 Maxwell electric distributed propulsion vehicle.

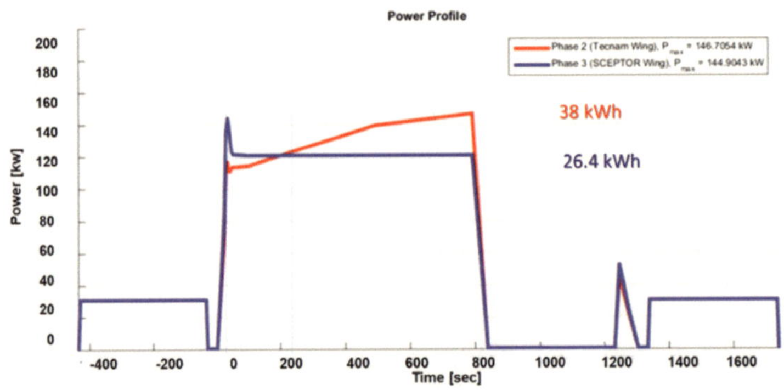

Figure 12.
X-57 Maxwell standard mission power requirement estimates.

(area under the power requirement curve). For example, during the cruise time interval of 800 seconds (0.22 hours), at constant power the energy required is $120 \times 0.22 = 26.4$ kWh with the X-57 wing (blue line). Based on the current mission analysis utilizing the original *Tecnam* wing, 38 kWh is required to meet the peak power demand of 145 kW (red line).

Assuming M-SHELLS could produce 1000 W/kg specific power at a 75 Wh/kg specific energy, a 120 kg M-SHELLS package would ideally provide 120 kW of power and 9 kWh of energy. Given the 120 kW of power required during cruise with the X-57 wing (blue line), the M-SHELLS package could supply energy for a duration of 0.075 hours, or 270 seconds, at level cruise.

A brief structural analysis of the fuselage was conducted, where a reinforced M-SHELLS multifunctional panel can be safely substituted to partially replace the lightly loaded aluminum floor structure. **Figure 13** shows an example of fuselage floor deflection and shear stress with the original floor replaced by a reinforced composite panel with the M-SHELLS core. The five-layer composite sandwich panel consisted of two 0.05 inch thermoplastic sheets for reinforcement and insulation on the outer faces, which were bonded to the two 0.002 inch aluminum sheets on the inner faces over the 1.0 inch deep M-SHELLS core. For this example, the total distributed floor load is 265 lb (120 kg) distributed over the forward fuselage floor area. The fuselage floor deflection is nominal and the majority of the shear stresses across all plies are generally within the allowable limits except at the end support areas, where local reinforcements will be needed.

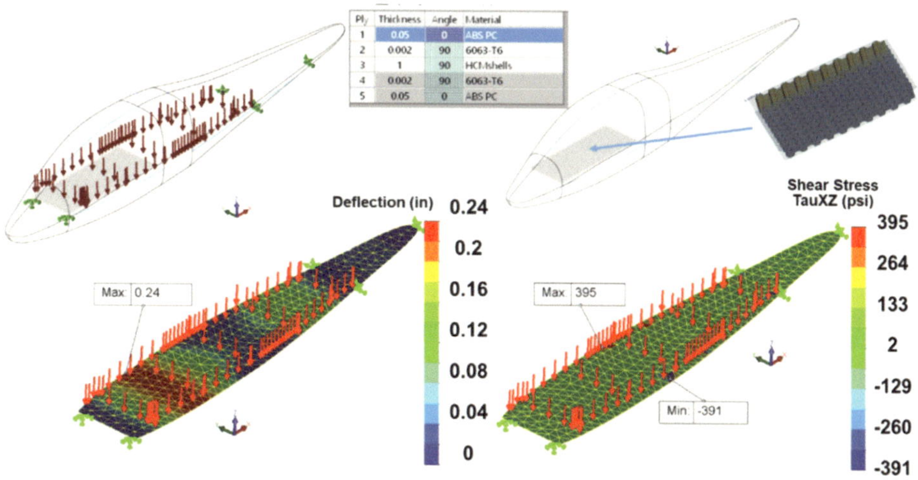

Figure 13.
X-57 floor deflection and shear stress analysis with 265 lb (120 kg) M-SHELLS distributed over the forward fuselage floor area.

5. Hybrid-electric aircraft

In the ARMD Advanced Air Transport Technology (AATT) project, several aircraft concepts are presently being studied to quantify the performance improvements and emissions reduction afforded by hybrid-electric propulsion. Jansen et al. [7] have conducted extensive systems analysis to evaluate the risks and benefits of a conversion from an all-fuel turbofan to a hybrid-electric turbofan engine concept. Among the propulsion options considered by this study, the "hFan" concept is a gas turbine-electric hybrid engine capable of operating in all-gas turbine, all-electric, or combined mode, depending on mission requirements. Conventional and truss-braced wing concepts with hybrid-electric propulsion were also investigated by Bradley and Droney [8, 9] at the Boeing Company.

Objectives of the NASA Electrified Aircraft Propulsion (EAP) research are to increase fuel efficiency and to reduce the emissions and noise levels of commercial transport aircraft. Primary EAP propulsion concepts include turboelectric, partially turboelectric, and hybrid-electric systems. Applications are presently being evaluated for regional jet and larger sized single-aisle aircraft. The overall goal is to demonstrate the viability of at least one of the EAP concepts. A hybrid-electric derivative of the N+3 technology conventional configuration (N3CC) is an ideal candidate for future applications of the M-SHELLS technology, by replacing lightly loaded portions of the fuselage structures where use of lightweight honeycomb panel is possible. The outer mold line (OML) of this aircraft concept [5] was developed using the Open Vehicle Sketch Pad tool [10, 11]. The internal structure of a fuselage segment of this vehicle was developed using SolidWorks [12] for finite element analysis. The structural analysis included a combination of aluminum and reinforced M-SHELLS composite panels for stress, deflection, and weight estimation. A block diagram of the FEM development and sizing process is presented in Appendix C.

Figure 14a shows the N3CC vehicle model with internal structure, and the detailed FEM of a fuselage segment is shown in **Figure 14b**. The fuselage section design loads consist of an internal cabin pressure of 18.4 psi, passenger floor load of 1 psi, and cargo floor load of 2 psi. The weight analysis of the N3CC hybrid concept fuselage segment with Al 7075-T6 construction is shown in **Table 2**. The total FEM weight of this all-aluminum fuselage segment is 4992 lb. This includes a passenger floor weight of 876 lb,

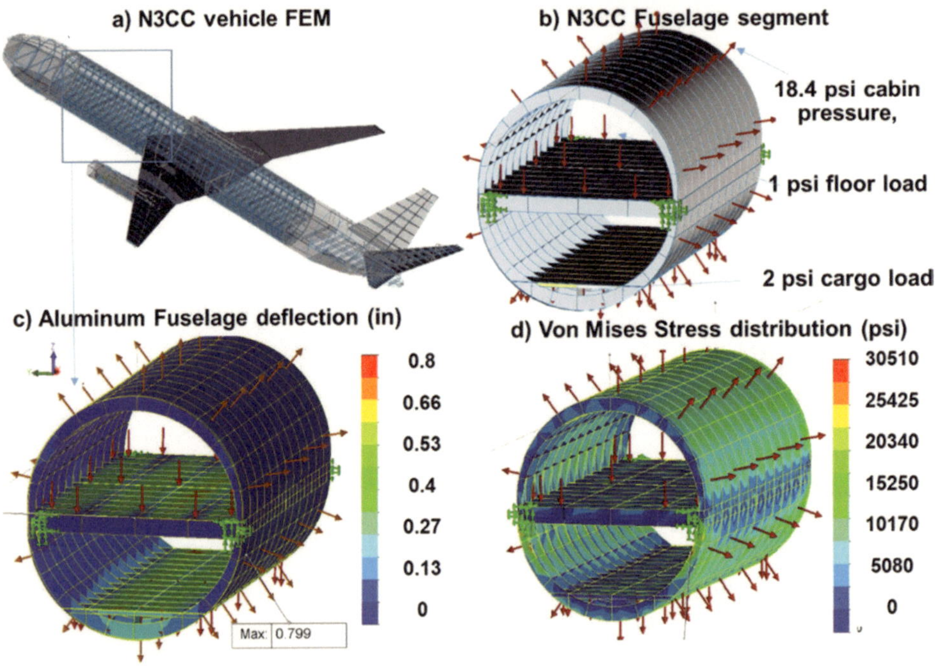

Figure 14.
N3CC fuselage segment analysis with aluminum 7075-T6 material construction.

Fuselage Assembly	height (in) 160		width (in) 148		length/segment 288 in			frame spacing 24 in		FEM		fuselage
Segment Structural Items	length	width	area	thickness	volume	material	density	item weight	no. of items	total weight		segment component
segment 1	inch	inch	in^2	in	in^3		lb/in^3	lb			unit	weight (lb)
fuselage frame (oval)	segment		3699	0.208	769	AL	0.1015	78	13	1015	lb	per segment
fuselage outer skin	288	493	141984	0.156	22150	AL	0.1015	2248	1	2248	lb	pass floor
passenger floor	288	148	42624	0.104	4433	AL	0.1015	450	1	450	lb	876
passenger floor beam	148	12	1776	0.104	185	AL	0.1015	19	13	244	lb	
passenger floor frame	288	12	3456	0.104	359	AL	0.1015	36	5	182	lb	outer shell
cargo floor skin	288	99	28512	0.104	2965	AL	0.1015	301	1	301	lb	3461
cargo floor frames	3	99	297	0.104	31	AL	0.1015	3	13	41	lb	Cargo floor
wing carry-thru beam	150	20	3000	0.104	312	AL	0.1015	32	2	63	lb	342
keel beam	288	19	5472	0.15	821	AL	0.1015	83	3	250	lb	keel+wing beam
longitudinal stringers	288	5	1440	0.104	150	AL	0.1015	15	13	198	lb	313
fuselage segment weight			295 sq ft passenger floor area					16.9 lb/sq ft		4992	lb	4992

Table 2.
Weight analysis of N3CC fuselage segment with aluminum 7075-T6 construction.

an outer shell weight of 3461 lb, a cargo floor weight of 342 lb, and the total keel-beam and cross-beam weight of 313 lb. **Figure 14c** shows the all-aluminum fuselage deflection and **Figure 14d** shows the von Mises stress distribution.

Figure 15 shows the modified fuselage section in which the passenger and cargo subfloor cross-beams were replaced with the five-layer reinforced composite panels with honeycomb core (5LCHC). The sandwich panels consisted of 1 inch deep M-SHELLS honeycomb core and 0.002 inch aluminum ply and 0.05 inch thermoplastic ply on each side. **Figure 15a** shows the N3CC fuselage model and design load. As before, the fuselage section design loads consisted of an internal cabin pressure of 18.4 psi, passenger floor load of 1 psi, and cargo floor load of 2 psi. The passenger subfloor and cargo subfloor cross-beams are now replaced with this five-layer bonded composite panel with M-SHELLS honeycomb core (**Figure 15b**). **Figure 15c** shows a significant increase in the maximum floor deflection compared to the all-aluminum construction shown in **Figure 14c**. **Figure 15d** shows maximum von

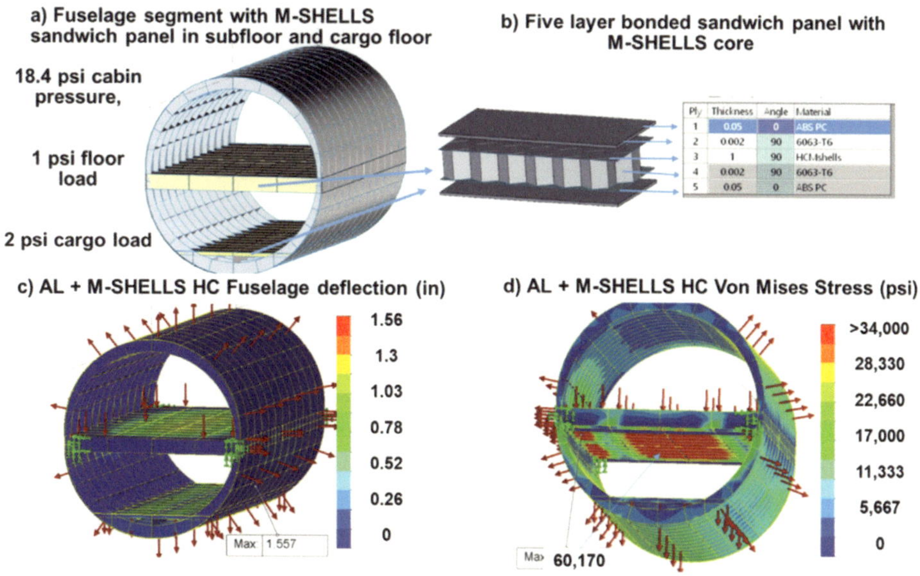

Figure 15.
N3CC fuselage segment analysis with passenger and cargo subfloor cross-beams replaced by reinforced composite panels with M-SHELLS core.

Fuselage Assembly	height (in)	160 width (in)		148 length/segment 288 in				frame spacing 24 in		FEM		fuselage
Segment Structural Item	length inch	width inch	area in²	thickness inch	material	property	unit	item weight lb	no. of items	total weight	unit	segment-1 component weight (lb)
fuselage frames (oval)	segment 1		3699	0.208	AL	0.1015	lb/in³	78	13	1015	lb	per segment
fuselage outer skin	288	493	141984	0.156	AL	0.1015	lb/in³	2248	1	2248	lb	pass. floor
passenger floor	288	148	42624	0.104	AL	0.1015	lb/in³	450	1	450	lb	728
passenger floor beam	148	12	1776	1.104	5LCHC	0.00689	lb/in²	12	13	159	lb	
passenger floor frame	288	12	3456	1.104	5LCHC	0.00689	lb/in²	24	5	119	lb	outer shell
cargo floor skin	288	99	28512	0.104	AL	0.1015	lb/in³	301	1	301	lb	3461
cargo floor frame	3	99	297	1.104	5LCHC	0.00689	lb/in²	2	13	27	lb	cargo floor
wing carry-thru beam	150	20	3000	0.104	AL	0.1015	lb/in³	32	2	63	lb	328
keel beam	288	19	5472	0.15	AL	0.1015	lb/in³	83	3	250	lb	keel+wing beam
longitudinal stringer	288	5	1440	0.104	AL	0.1015	lb/in³	15	13	198	lb	313
one fuselage segment weight			295 sq ft passenger floor area					16.37 lb/sq ft		4830	lb	4830

Table 3.
Weight analysis of N3CC fuselage segment with aluminum 7075-T6 and M-SHELLS honeycomb composite panel.

Mises stress distribution across all ply, which are significantly higher locally in the passenger sub-floor cross-beam.

The weight analysis of the N3CC hybrid concept fuselage segment with aluminum and M-SHELLS composite panels is shown in **Table 3**. The total FEM weight of this fuselage segment is 4830 lb. The passenger floor weight is reduced to 728 lb from 876 lb for the previous case. The aluminum outer shell weight remains 3461 lb. The cargo floor weight is reduced to 328 lb from 342 lb. The total keel-beam and cross-beam weight remains 313 lb. Thus, the weight reduction for one fuselage segment is 162 lb or 3.2%, at the cost of higher fuselage deflection and stress, but without risking the structural integrity (**Figure 15c** and **d**).

Since this substitution resulted in large increases in deflection and stress in the passenger floor (**Figure 15c** and **d**), additional sub-floor support in the cargo hold area was examined as shown in **Figure 16a** and **b**. The corresponding structural deflection and stress distribution are shown in **Figure 16c** and **d**. The maximum deflection was reduced significantly and the von Mises stress distributions were within the allowable limits. The additional M-SHELLS weight was 173.5 lb. Hence, the net weight increase was 11.5 lb (0.3%) per segment, compared to all aluminum construction, while adding 56 cubic foot of M-SHELLS storage volume. The fuselage section weight comparison summary from the three designs is presented in **Figure 17**.

These weight calculations with the reinforced M-SHELLS panel did not include copper current collectors, separator layers, and electrolyte that are required to complete the energy storage functionality but do not add to the structural strength. Appendix B shows the M-SHELLS panel density and properties. A full vehicle structural and systems analysis for the N3CC derivative with hybrid-electric propulsion was presented by Olson and Ozoroski [2] to predict the multifunctional performance and weight benefits with higher specific energy M-SHELLS replacing major primary structure. Their study showed that by offsetting the weight of some of the vehicle's primary batteries or mission fuel, an overall weight savings can be achieved through multifunctionality. An initial version of the paper was proposed for presentation in [13].

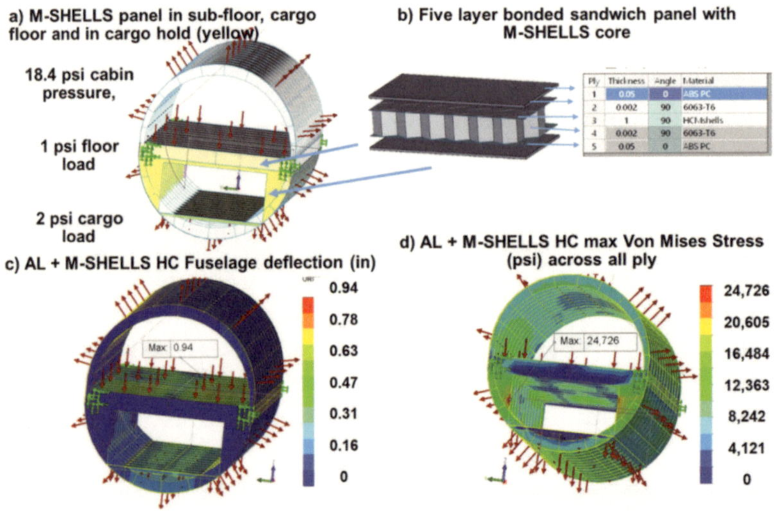

Figure 16.
N3CC fuselage segment analysis with additional reinforced M-SHELLS panel added to the subfloor cargo area.

fuselage segment Component	ALL AL fuselage	AL + M-SHELL sub-floor	AL + M-SHELL sub-floor and Cargo area	unit
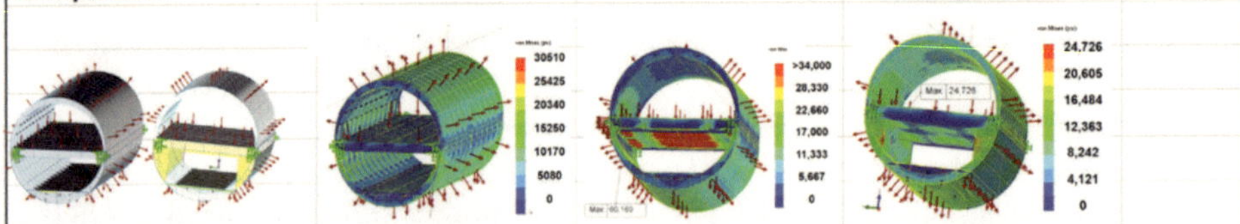				
passenger floor	876	728	728	lb
fuselage outer shell	3461	3461	3461	lb
Cargo floor+ stiffener	342	328	501	lb
keel+wing cross beam	313	313	313	lb
Total wt /segment	4992	4830	5003	lb
% change in weight		-3.2%	0.23%	

Figure 17.
Summary of weight comparison from the three fuselage segment design.

6. Concluding remarks

The Multifunctional Structures for High Energy Lightweight Load-bearing Storage (M-SHELLS) research project is described. The proposed project goals were to develop M-SHELLS in the form of honeycomb coupons and subcomponents, integrate them into the structure, and conduct low-risk flight tests onboard a remotely piloted small aircraft. The M-SHELLS sample units were scheduled for flight testing onboard a remotely piloted small aircraft named *Tempest*. Detailed finite element models of this small test aircraft were developed for basic structural strength and accurate weight analysis. The *Tempest* wing FEM was refined to include the unique wing construction and provide a closer match with the wing deflection results from a bench test. The component weight analysis from the finite element analysis and load test data were correlated. Finite element analysis results of *Tempest* with a reinforced five-layer M-SHELLS composite panel replacing the mid-fuselage floor were presented. Approximately, 2.2 lb of M-SHELLS would provide power for 10 minutes of cruise flight. Although the planned flight test was cancelled due to the project constraints, the analysis results indicate that the mid-fuselage floor composite multifunctional panel could provide both structural integrity and electrical energy to supplement the existing battery.

The NASA X-57 Maxwell distributed electric propulsion test vehicle was used as an example for potential application of the M-SHELLS technology. The fuselage floor structure was selected for substituting a reinforced composite panel with M-SHELLS core. A structural analysis of the fuselage floor indicated that it could self-support a 265 lb (120 kg) M-SHELLS system, providing sufficient power and energy for 270 seconds of cruise flight. The fuselage floor deflection is nominal and the majority of the shear stresses are generally within the allowable limits. For future applications of M-SHELLS, structural analysis of an advanced transport aircraft fuselage segment is presented. Secondary aluminum structure in the fuselage sub-floor and cargo area were replaced with reinforced composite panels with M-SHELLS honeycomb core. Fuselage structural analyses associated with three cases were described. The weight estimation with the reinforced composite M-SHELLS panels replacing the passenger sub-floor indicated a 3.2% reduction in fuselage weight, at the cost of higher deflection and stresses, but without risking the structural integrity. With additional M-SHELLS panels in the cargo hold area, the deflection and stresses were reduced. But, the net weight of the fuselage segment increased by 11.5 lb (0.3%) compared to all aluminum construction, while adding 56 cubic foot of M-SHELLS volume and ~22 kWh of energy capacity/segment. These weight calculations were with the reinforced M-SHELLS panel with 11.9 lb/ft^3 density. This calculation did not include reactive materials that are required to complete the energy storage functionality.

Acknowledgements

The authors thank the NASA Transformative Aeronautics Concepts Program (TACP), directed by Dr. John Cavolowsky for funding this research. The M-SHELLS sub-project is part of the Convergent Aeronautical Solutions (CAS) project under the TACP Program. CAS is directed by Isaac Lopez, Project Manager, Marty Waszak, Deputy Project Manager and Daniel Williams, CAS Liaison Officer at Langley. We thank Dr. Nicholas Borer, Jeff Viken, and the X-57 Maxwell Systems Analysis and Integration team for their guidance. Thanks are also due to Kevin Roscoe, Greg Howland, David A. Hare, Luke Laub, David Klassman, and Frank Leone for performing the *Tempest* structural weight measurement and deflection tests, and to Russell Smith and Brett Bednarcyk for leading the fabrication team, and for providing the M-SHELLS coupon structural test results. Authors also acknowledge technical support from Dr. Dave Krause and Dr. Patricia Loyselle, NASA Glenn Research Center, William Kimmel, Chief Technologist, Systems Analysis and Concepts Directorate (SACD), Philip Arcara, Branch Head, and Mark Guynn, Assistant Branch Head, Aeronautics Systems Analysis Branch (ASAB). Thanks are also due to all the technical reviewers for improving the report significantly. Technical discussions with Dr. Wu Li, Dr. Natalia Alexandrov, Jason Kirk and Dr. Jesse Quinlan for improving the chapter are greatly appreciated.

Appendix A. Material elastic property and density

Material property	Advanced stitched composite	AL 7075 T6	ABS Thermoplastic	M-SHELLS test specimen	unit
Elastic Modulus (T)	9750000	10442710	290075	11188	psi
Poisson's ratio	0.4	0.33	0.394	0.39	
Shear mudulus	2570000	3901515	46250	4024	psi
Mass density	0.0526	0.0101	0.04	0.0039	lb/in^3
Tensile strength	105100	82760	4350	90	psi
Compressive strength	79200	80000	4000	120	psi
Yield strength	46500	73244	4355	90	psi

Appendix B. Sandwich panel density

Sandwich Panel	Description	cell	density	density	unit
Hexcel Sandwich	AL-Hexcel HC	1/4 5052-0.002	4.3 lb/ft^3=	0.0025	lb/in^3
Hexcel Sandwich	AL-Hexcel HC	1/4 5052-0.004	7.9 lb/ft^3=	0.0046	lb/in^3
5 layer bonded MSHC	5LMSHC (0.002)	core 1/4 5052-0.002	11.9 lb/ft^3=	0.0069	lb/in^3
5 layer bonded MSHC	5LMSHC (0.004)	core 1/4 5052-0.004	16.19 lb/ft^3 =	0.0094	lb/in^3

Appendix C. Structural model development and analysis process

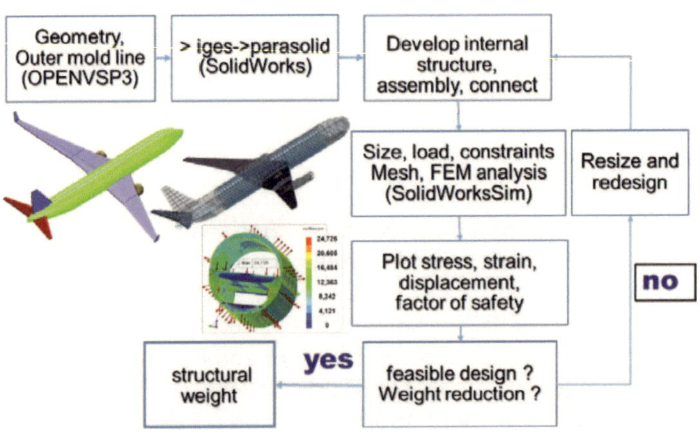

Author details

Vivek Mukhopadhyay
NASA Langley Research Center, Hampton, Virginia, United States

*Address all correspondence to: v.mukhopadhyay-1@nasa.gov

References

[1] Krause D, Loyselle P. Multifunctional structures with energy storage—Enabling hybrid electric propulsion for commercial aircraft by melding load-carrying structure with energy storage to save weight uniting expertise at AFRC, ARC, GRC, and LaRC. ARMD, Convergent Aeronautical Solution (CAS) Proposal. 2016

[2] Olson ED, Ozoroski TA. System-level benefits of multifunctional structure/energy storage concepts for an advanced hybrid-electric commercial aircraft. In: Systems Analysis and Concepts Directorate Technical Forum and NASA Technical Report. 2018

[3] Newman T. VAPOR—NASA University design challenge, 2014-2015, aerospace systems design [MS thesis]. Aerospace Engineering Department, Georgia Institute of Technology; 2015

[4] Borer KN. Engineering Aviation's Next Transformation: Design Considerations for Electric Aircraft. NASA Langley Aeronautics Systems Analysis Branch Technical Forum. 2015

[5] Mukhopadhyay V, Ozoroski TA, McMillin MA. Structural configuration analysis of advanced flight vehicle concepts with distributed hybrid-electric propulsion. In: AIAA Sci-Tech 2018 (AIAA 2018-1747)

[6] Clarke S. SCEPTOR power system design: Experimental electric propulsion system design and qualification for crewed flight testing. In: AIAA Aviation Technology, Integration, and Operations Conference, NASA/NTRS document 20160007822; 13-17 June 2016; Washington, DC, United States

[7] Jansen RH, Bowman C, Jankovsky A, Dyson R, Felder J. Overview of NASA Electrified Aircraft Propulsion Research for Large Subsonic Transports. NASA/NTRS document 20170006235, Washington, DC. Cleveland, Ohio: NASA Glenn Research Center; 2017

[8] Bradley MK, Droney CK. Subsonic Ultra Green Aircraft Research: Phase I Final Report. NASA CR-2011-216847; Long Beach, CA: The Boeing Company; 2011

[9] Bradley MK, Droney CK. Subsonic Ultra Green Aircraft Research: Phase II—Volume II—Hybrid Electric Design-Exploration. NASA CR–2015-218704/Volume II; Long Beach, CA: The Boeing Company; 2015

[10] Gloudemans JR, McDonald R. Improved geometry modeling for high fidelity parametric design. In: 48th AIAA Aerospace Sciences Meeting and Exhibit; AIAA-2010-0659

[11] Hahn A. Vehicle sketch pad: Parametric geometry for conceptual aircraft design. In: 48th AIAA Aerospace Sciences Meeting, Orlando, FL. 2010 (AIAA-2010-657)

[12] SolidWorks and SolidWorks Simulation User Manual. SolidWorks Corporation. Waltham, Massachusetts: Dassault Systèmes; 2017

[13] Mukhopadhyay V, Olson ED, Ozoroski TA. Structural analysis of test flight vehicles for application of multifunctional energy storage system. In: AIAA SciTech Forum, San Diego, CA. 2019 (AIAA-2019-0551)

Chapter 2

Sustainable and Efficient Hydroforming of Aerospace Composite Structures

Bo C. Jin, Xiaochen Li, Karl Neidert and Michael Ellis

Abstract

Hydroforming, in comparison with sheet stamping, is an efficient and economical manufacturing process for complex-shape aerospace composite parts because it does not require the use of a female die. The hydroforming manufacturing method is expected to greatly increase the formability of composite parts by using a controllable heated and pressurized fluid that acts as a support for the composite sheet throughout the forming process. The design of a hydroforming process and a machine to shape complex aerospace composite parts is proposed in this chapter. The design and analysis of a sheet metal hydroforming machine with composite overwrap are presented to sustainably and efficiently produce not only the aerospace composites but also dual-phase and bake hardened steel parts with complex 3D geometry.

Keywords: aerospace composite structures, sustainable manufacturing, hydroforming

1. Introduction

1.1 Background

Sheet hydroforming is a process that was primarily developed for the needs of the aircraft and aerospace industry. In sheet hydroforming, formed tooling blocks are placed in the loading tray of a pressure vessel, and pre-cut sheet metal blanks are placed over the blocks. Throw pads are then placed over the blanks to cushion sharp edges. The tray is then fed into the pressing chamber as a thick elastic blanket is unrolled over the tool and sheet metal. It is then backfill pressurized with hydraulic fluid under ultra-high pressure. The elastic fluid cell blanket diaphragm expands and flows downward over and around the metal blank. The sheet metal is pressed to follow the contour of the die block, exerting an even, positive pressure at all contact points. As a result, the metal blank is literally wrapped to the exact shape of the die block. The press is then depressurized for unloading the tray [1]. This process is ideal for prototyping and low volume production in aluminum, titanium, stainless steel, and other malleable aerospace alloys such as metal-composite panels in low volumes.

The primary pressure containment vessels used in these machines are designed to contain ultra-high pressures. In some cases, the internal pressures can be as high as 137.90 MPa (20,000 psi). In small diameter tubing this is a notable pressure.

 IntechOpen

However, as the diameter and area of the pressure chamber increases, the total pressure is applied to a significantly larger surface area, resulting in total loading forces that are exceedingly high.

Current designs in operation feature a traditional circular cross-section, a familiar geometry to most designers. However, the actual pressures exerted inside the chamber are not cylindrically uniform, as they would be in a fluid or gas pressure tank. Instead, the pressure loading originates from hydraulic working fluid inside a fundamentally rectangular shaped tray. These loads are then transferred into solid metal blocks called yoke plates that in turn each press against the walls of the pressure containment chamber with varying degrees of force.

The non-uniform pressure loading of the cylindrical pressure vessel wall results in localized forces that cause engineers to use excessive or unnecessary material, which increases both the weight and cost of the equipment **Figure 1**.

The objective of this project was to optimize the cross-sectional profile, attributes, and material usage of a pressure containment vessel for use in a hydroforming manufacturing press. The new design was to be preferable in cost, weight and overall performance. Finite element analysis was used extensively to validate the current operational design, material alternatives and the optimized cross-section designs proposed. The design is a modular construction consisting of several pressure containment sleeve rings fabricated from layers of radially axial wound high strength composite fiber filament infused with resin stabilizer over a metal liner. Into this envelope of several joined compression rings slides a movable pressure vessel that features a top-load clamshell cartridge type design. Integral to the lid of the forming chamber is a series of elastomeric tubes that work in unison to produce a type of a high-pressure hydroforming diaphragm. This modular "sleeve over sandwich" pressure containment scheme is designed to enable the system to be easily configured in various shapes, sizes and lengths. It is also conceived to improve functionality, capability and serviceability. Because of the unique properties of the design, it can be easily configured in various lengths so that a wide range of products can be produced including 100 kW wind turbine blades.

The proposed concept sets forth numerous innovative breakthroughs that infuse legacy hydroforming technology with renewed vigor and greatly improved competence. The design is conceived to deliver enhanced functionality, capability, cost and serviceability as well as resale value. State-of-the-art sensing and computation enable many of the advancements. The parametric geometrical modeling of the liner and composite overwrap was performed using FEA software [2] which is proved to be an

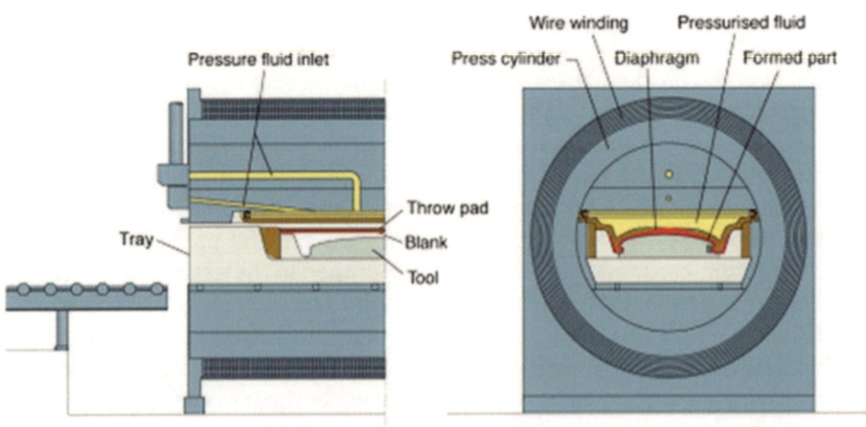

Figure 1.
Typical configuration of current commercial solutions (courtesy of Avure Inc.).

efficient method to verify the design and optimize geometry of advanced composite structures [3–10]. Parameters such as composite overwrap winding thickness, and the geometric outline of the liner and containment, were parametrically investigated to obtain optimized stress–strain relationships under hyper-pressure.

1.2 Objectives and structure

Currently the most powerful hydroforming pressure forming chambers are cylindrical constructions of high-strength, pre-tensioned steel wire wound over solid steel winding armatures. They are designed to meet "leak-rather-than-break" criteria". Some frames of this type can contain forces up to 137.89 MPa (20,000 psi) of operational pressure.

The objective of this project regarding the pressure containment system is to develop a non-cylindrical master section design that is comprised fundamentally of three elements: (a) Composite windings, b) a winding core, c() aluminum yoke plates. The design failure target for pressure containment is 165.47 MPa (24,000 psi). That is 137.89 MPa (20,000 psi) with a 20% safety factor or 110.32 MPa (16,000 psi) with a 50% safety factor. The containment section is intended to be configured to reduce weight significantly over "HS Steel Over Steel" chamber construction. It is intended to allow a completed machine to rest directly on a standard factory floor without additional floor structure reinforcements. The assumption for the proposed cylinder construction is composite over aluminum yoke plates. Composites may include glass, Kevlar 49, and carbon fibers. This study uses a carbon fiber source (Zoltek Panex 35 Continuous Tow pre-preg). The machine is meant to operate under cold shell start up conditions to 50% polymer plastic point. And the chamber operating temperature is assumed to be –17.78°C (0°F) to 48.89°C (120°F).

Aluminum alloys considered are 6061-T6 and 7075-T6. Windings and press frame are assumed for this study to be of matched metal type. Compressive strength of concrete for resting footprint loading is set at 17.24 MPa (2500 psi). The mechanism of operation of the machine assumes that pressure is applied by the injection of working fluid into a forming chamber cassette that has been loaded into the void area. The internal forming chamber cassette is not included in this modeling study. Pressure will be applied at full pressure to the side and horizontal walls of the inner yoke plate areas. The chamber has a fundamentally uniform cross section that is suited to sectional analysis. Loads will propagate from the inner void, into the yoke plates, into the winding frame and ultimately into the composite winding bobbin. Maximum vertical deflection at full pressure load of 165.47 MPa (24,000 psi) is assumed to be 6 mm.

If possible, it is desired that the design be low profile in appearance, resembling that of a toroidal ellipse section. This is desired to allow the installation and use of the machine without the addition of false load floors to support ergonomic reach over heights on larger than 1524 mm (60 in) wide forming cartridges. It is also assumed that a cosmetic outer cover may be applied to the cylinder design. This cosmetic cover will not be included in this design exploration. The project will begin with a baseline developed from a replication of a purely cylindrical design: "A". The effects of the loading properties will be applied to additional non-cylindrical designs B–E. The purpose of these designs is to build conceptual understanding by exploring radically unique cross-section designs. Based on these findings, special consideration is to be given to develop additional sections that may produce a design of reduced weight and cross-sectional height vs. a purely cylindrical design. The effects of steel vs. aluminum yoke plates are also to be investigated to compare strength to weight to size performance.

In Section 2 of this paper, we introduce the optimal design procedure and the design configurations A–E, and E1–E4. Sections 3 and 4 describe the results and discussion, and future perspective and opportunities.

2. Optimal design procedure

2.1 Basic structural configuration and FEA design parameters

Any composite overwrapped pressure vessel (COPV) design requires an analysis of the liner, the fiber overwrap, and the interaction between the two. COPV liners may be of ductile materials with only minimal load-sharing capabilities, such as ductile metal composites [11–19], glass fiber reinforced angle interlock composites [20–22], or carbon fiber reinforced composite oriented strand boards [5, 23–29].

The general layout and components of the baseline cross-section design (Section "A") are presented in **Figure 2a**. The outer diameter of Section "A" is roughly 60 cm. The design variables for the optimization of the circular cross-section "A" are presented in **Figure 2b**. These design variables are modified systematically throughout the design exploration procedure, allowing for different evolving design configurations.

In the same manner, four different probe locations were defined (**Figure 2c**) to obtain stress and displacement results of the tested cross-section designs and allow a quantitative comparison of the effects of each individual design variable, their respective interactions and the overall behavior of the different design configurations. These probing locations will be used for all the analysis carried out in order to obtain consistent results.

2.2 Yoke plate cutting angle concept

A new design concept referred to as "Yoke plate cutting angle" is introduced as a design variable. This unique concept provides control over the stress distribution at the yoke plates interface by means of a resulting friction force generated at the surface-to-surface contact interface. The angled (non-horizontal) contact interface between the yoke plates and side yoke plates allows both plates to work in unison to carry the pressure load while still providing some accommodation for stress deflection by allowing slip plane movement between the plates. The yoke plates' cutting

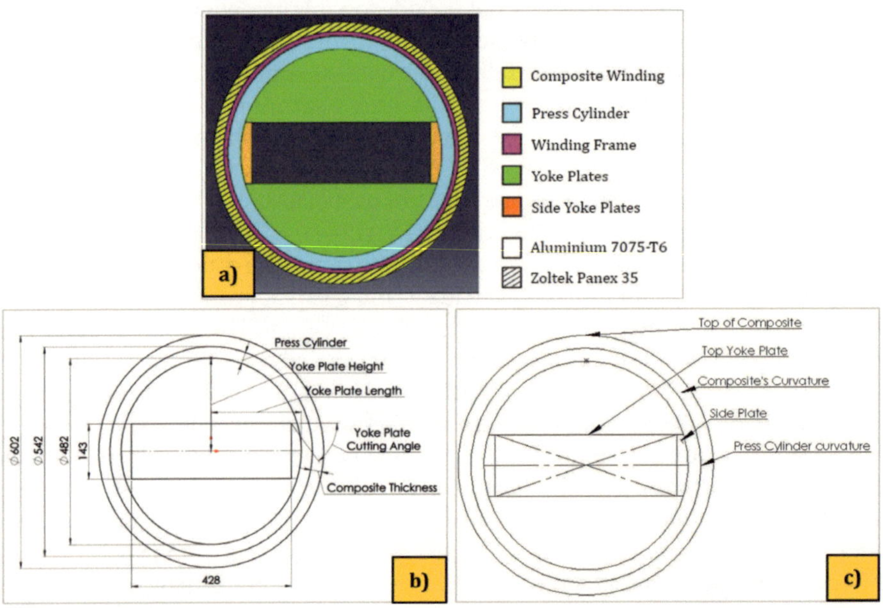

Figure 2.
(a) General component layout and materials used, (b) general dimensions and design parameters and (c) stress-probing locations.

Elastic modulus	E	71.7 GPa
Poisson	Nu	.33
Density	Rho	2.81e−9
Yield point	σ	503 MPa

Table 1.
Material properties of winding frame and yoke plates. (7075-T6 aluminum).

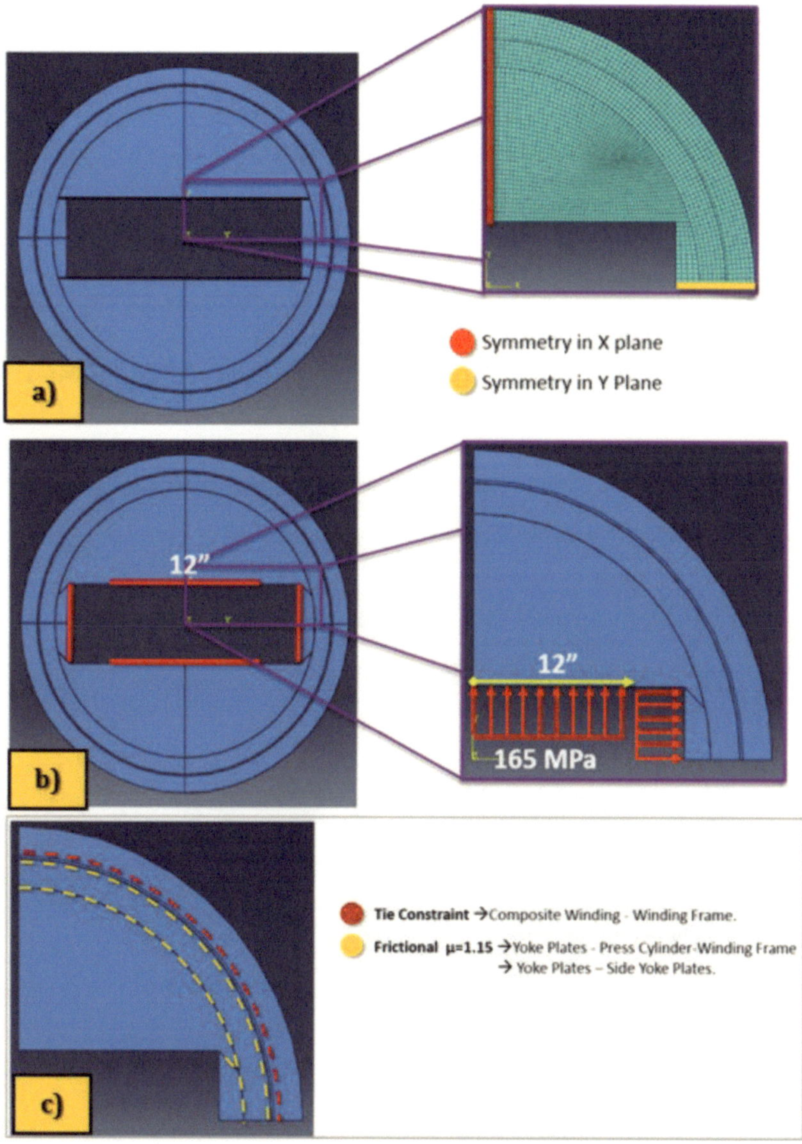

Figure 3.
(a) Boundary conditions and mesh, (b) applied loads and (c) contact definitions.

angle dictates the magnitude of the frictional force occurring at the interface. Thus, special attention must be paid to this dimension. A precise yoke plate cutting angle definition will allow contact related stresses to be just below the material's yield, while helping reduce the stress of the composite winding. Coulomb dry friction is a good starting point to approximate the maximum cutting angle that generates a "stick" condition (no sliding) between the two surfaces. However, if no sliding occurs, stress concentration around the contact interface is bound to rise past the allowable limit—thus, some sliding must be ensured to provide stress relief.

2.3 General finite element model setup

2.3.1 Geometry and discretization

All the cross sections to be analyzed are expected to be symmetric in both horizontal and vertical directions due to the homogenous pressure that they are subjected to. The finite element models constructed, take full advantage of this geometric symmetry by simulating only one quarter of each cross section. This reduces computational time significantly **Table 1**.

Solid 3D elements with an approximate global size of 2.5 mm were used to create a fine mesh representative of the geometric features (**Figure 3a**).

2.3.2 Materials

The following material properties were applied to the aluminum yoke plates and winding frame:

For the composite winding, carbon fiber prepreg (Zoltek Panex 35) was used. This material's properties are presented in **Table 2**.

2.3.3 Boundary conditions

Symmetry boundary conditions were applied as shown in **Table 3**.

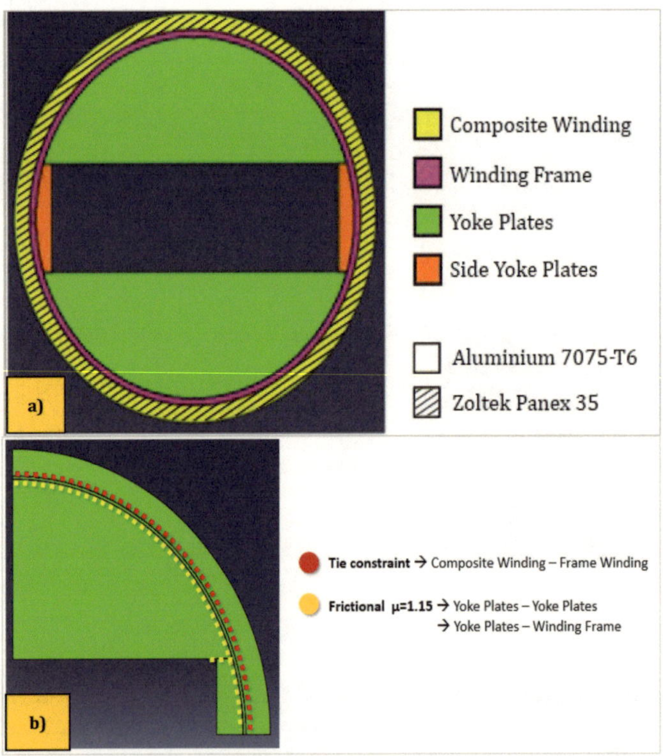

Figure 4.
(a) Updated cross-section assembly without press cylinder tube and (b) updated contact definition.

2.3.4 Loading conditions

A homogenous pressure of 165 MPa is applied to the inner faces of the yoke plates as shown in **Figure 3b**.

2.3.5 Contact definition

Frictional contacts were defined between the yoke plates and the winding frame (**Figure 3c**). A tie constraint was used to bond the winding frame and the composite winding, as it is assumed that no movement occurs between these two components.

2.3.6 Section "A" test matrix

Section "A" is a pure circular cross section imitative of that of commercially available hydroforming machines with similar operating pressure capabilities. However, these dimensions must be optimized for the use of composite winding instead of the traditional pre-tensioned steel wire.

The test matrix for the Section "A" optimization is presented in **Table 4**. The non-horizontal angled cut contact interface between yoke plates is introduced as an additional design variable with the intention of reducing yoke plates stress, while the press cylinder thickness is reduced until it is completely removed.

2.3.7 Updated finite element model without press cylinder

Following the results obtained from the analysis of section "A," the finite element model assembly was modified. The new assembly is essentially the same as the original, with the exception that the press cylinder has been removed, and the yoke plates were extended to fill out the space of the press cylinder. The same materials are used for each component. Likewise, the same symmetry conditions and load cases are applied.

Elastic modulus 1	E1	134 GPa
Elastic modulus 2 & 3	E2, E3	129 GPa
Poisson ratio	Nu12, Nu13, Nu23	0.34
Shear modulus	G	4.84 GPa
Density	Rho	2.81e−9
Failure stress	σ	1903 MPa

Table 2.
Composite winding material properties (Zoltek Panex 35 carbon Prepreg [30]).

Symmetry plane	Degrees of freedom					
	Translation			Rotation		
	Tx	Ty	Tz	Rx	Ry	Rz
X	0	Free	Free	Free	0	0
Y	Free	0	Free	0	Free	0

Table 3.
Symmetry boundary conditions.

ID #	Composite thickness	Press cylinder thickness	Yoke plate radius	Plate cutting angle
1	30	30	241	0
2	30	30	241	−45
3	30	40	241	0
4	30	30	241	45
5	30	30	241	90
6	30	30	251	45
7	30	30	251	−45
8	30	10	261	−45
9	25	0	261	−45, rounded corners
10	25	0	251	−45, rounded corners
11	25	0	241	−35, rounded corners
12	25	0	241	0
13	25	0	241	0, rounded corner

Table 4.
Section "A" design configurations test matrix.

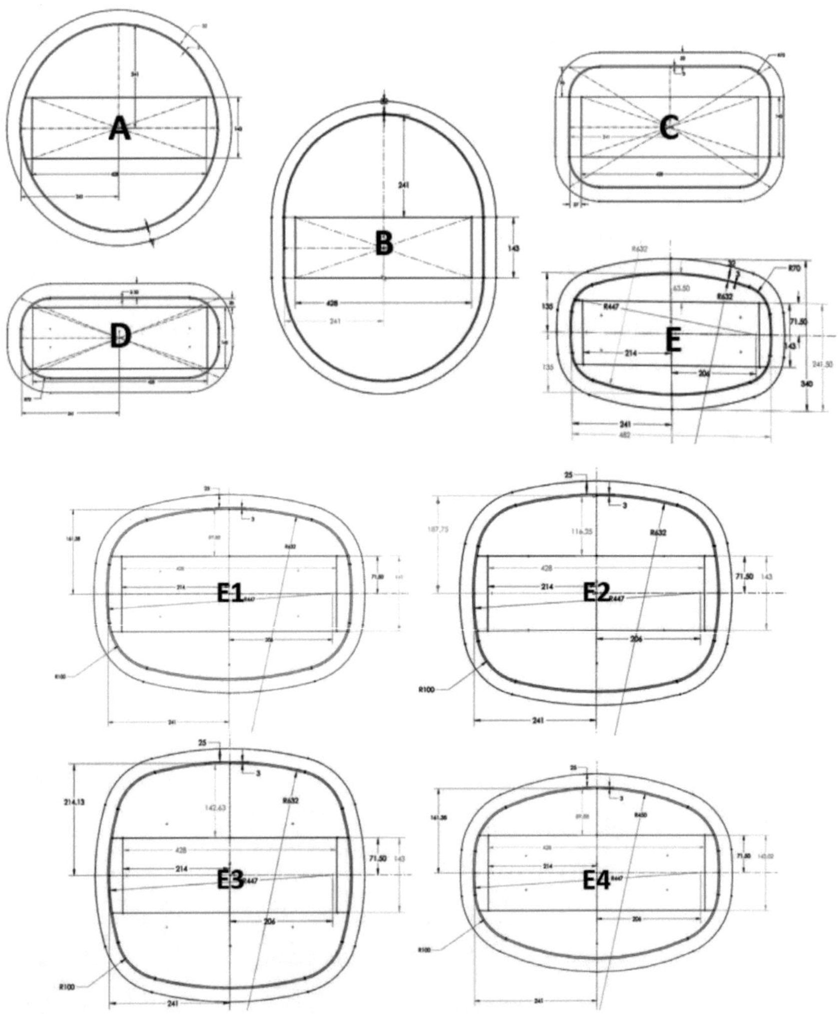

Figure 5.
Proposed cross section shapes A, B, C, D, E and E variants.

2.3.8 Updated contact definition

Without the press cylinder component, the yoke plates are in direct contact with the winding frame, so a new frictional contact was added between these two parts, shown in **Figure 4b**.

2.4 Finite element models for sections "B," "C," "D," and "E"

A similar finite element model and analysis was carried out for the rest of the cross-section designs shown in **Figure 5**. This design aims to determine the effect of the reduced overall cross section height, and different curvatures applied to the main shape of the cross section. However, due to the removal of the press cylinder, the subsequent models were readjusted as previously shown in 2.3.7 & 2.3.8.

2.5 Variations of section "E"

In an effort to improve the performance of the promising section "E", additional designs derived from section "E" were analyzed (**Figure 5**). Moreover, a design matrix (**Table 5**) was constructed based on defined design parameters: composite

		25mm Composite Winding						35mm Composite Winding								
		Aluminium			Steel			Aluminium			Steel					
		Corner Radius			Corner Radius			Corner Radius			Corner Radius			Max. Stress (MPa.)		Displacement
Design #	Section	0 mm.	10 mm.	20 mm.	0 mm.	10 mm.	20 mm.	0 mm.	10 mm.	20 mm.	0 mm.	10 mm.	20 mm.	Composite	Yoke Plate	Yoke Plate
25	E1	1	0	0	0	0	0	0	0	0	0	0	0	2038.0	4534.0	11.3
26	E1	0	1	0	0	0	0	0	0	0	0	0	0	2004.0	4377.0	11.7
27	E1	0	0	1	0	0	0	0	0	0	0	0	0	1949.0	3375.0	12.0
28	E1	0	0	0	1	0	0	0	0	0	0	0	0	1269.0	6701.0	6.6
29	E1	0	0	0	0	1	0	0	0	0	0	0	0	1234.0	6516.0	6.8
30	E1	0	0	0	0	0	1	0	0	0	0	0	0	1178.0	4938.0	6.8
31	E1	0	0	0	0	0	0	1	0	0	0	0	0	1668.0	3982.0	9.4
32	E1	0	0	0	0	0	0	0	1	0	0	0	0	1648.0	3864.0	9.7
33	E1	0	0	0	0	0	0	0	0	1	0	0	0	1620.0	2979.0	10.0
34	E1	0	0	0	0	0	0	0	0	0	1	0	0	1102.0	6078.0	5.8
35	E1	0	0	0	0	0	0	0	0	0	0	1	0	1079.0	5930.0	5.9
36	E1	0	0	0	0	0	0	0	0	0	0	0	1	1038.0	4515.0	6.0
37	E2	1	0	0	0	0	0	0	0	0	0	0	0	1393.0	3642.0	6.3
38	E2	0	1	0	0	0	0	0	0	0	0	0	0	1390.0	3666.0	6.4
39	E2	0	0	1	0	0	0	0	0	0	0	0	0	1372.0	2866.0	6.4
40	E2	0	0	0	1	0	0	0	0	0	0	0	0	815.9	5269.0	3.3
41	E2	0	0	0	0	1	0	0	0	0	0	0	0	805.7	5206.0	3.3
42	E2	0	0	0	0	0	1	0	0	0	0	0	0	792.0	3981.0	3.3
43	E2	0	0	0	0	0	0	1	0	0	0	0	0	1155.0	3182.0	5.5
44	E2	0	0	0	0	0	0	0	1	0	0	0	0	1151.0	3225.0	5.5
45	E2	0	0	0	0	0	0	0	0	1	0	0	0	1134.0	2534.0	5.5
46	E2	0	0	0	0	0	0	0	0	0	1	0	0	707.2	4833.0	3.0
47	E2	0	0	0	0	0	0	0	0	0	0	1	0	697.4	4818.0	3.0
48	E2	0	0	0	0	0	0	0	0	0	0	0	1	684.4	3703.0	3.0
49	E3	1	0	0	0	0	0	0	0	0	0	0	0	1009.0	3096.0	4.2
50	E3	0	1	0	0	0	0	0	0	0	0	0	0	1005.0	3309.0	4.2
51	E3	0	0	1	0	0	0	0	0	0	0	0	0	997.9	2279.0	4.3
52	E3	0	0	0	1	0	0	0	0	0	0	0	0	542.6	4488.0	2.1
53	E3	0	0	0	0	1	0	0	0	0	0	0	0	538.7	4709.0	2.1
54	E3	0	0	0	0	0	1	0	0	0	0	0	0	534.3	3196.0	2.1
55	E3	0	0	0	0	0	0	1	0	0	0	0	0	849.8	2710.0	3.7
56	E3	0	0	0	0	0	0	0	1	0	0	0	0	845.3	2923.0	3.7
57	E3	0	0	0	0	0	0	0	0	1	0	0	0	838.2	2021.0	3.7
58	E3	0	0	0	0	0	0	0	0	0	1	0	0	484.2	4132.0	1.9
59	E3	0	0	0	0	0	0	0	0	0	0	1	0	479.4	4374.0	1.9
60	E3	0	0	0	0	0	0	0	0	0	0	0	1	473.8	2979.0	1.9
61	E4	1	0	0	0	0	0	0	0	0	0	0	0	2033.0	5196.0	11.9
62	E4	0	1	0	0	0	0	0	0	0	0	0	0	2019.0	4792.0	12.4
63	E4	0	0	1	0	0	0	0	0	0	0	0	0	1970.0	3705.0	12.8
64	E4	0	0	0	1	0	0	0	0	0	0	0	0	1347.0	8145.0	7.2
65	E4	0	0	0	0	1	0	0	0	0	0	0	0	1308.0	7519.0	7.4
66	E4	0	0	0	0	0	1	0	0	0	0	0	0	1242.0	5687.0	7.5
67	E4	0	0	0	0	0	0	1	0	0	0	0	0	1659.0	4515.0	9.9
68	E4	0	0	0	0	0	0	0	1	0	0	0	0	1642.0	4173.0	10.2
69	E4	0	0	0	0	0	0	0	0	1	0	0	0	1613.0	3223.0	10.6
70	E4	0	0	0	0	0	0	0	0	0	1	0	0	1153.0	7285.0	6.3
71	E4	0	0	0	0	0	0	0	0	0	0	1	0	1120.0	6741.0	6.4
72	E4	0	0	0	0	0	0	0	0	0	0	0	1	1075.0	5125.0	6.5

*Red cells are above the material's Yield/Strength Point

Table 5.
Section E variations test matrix.

ID	Design variables					Results (MPa)				
	Composite thickness	Plate height	Plate length	L/H ratio	Plate cutting angle	Composite stress at top	Composite stress at curvature	Top plate stress	Side plate stress	Displacement (mm)
13	25	241	241	1	0	665	1807	355	383	2.98
14	25	200	250	1.25	0	1052	2002	587	432	5.21
15	25	200	250	1.25	0	1098	1983	628	538	4.93
16	25	200	300	1.5	0	1434	2173	852	457	8.38
17	25	200	300	1.5	−35	1395	2018	843	777	8.38
18	25	220	242	1.1	45	1428	2268	990	423	9.2
19	25	220	242	1.1	0	791	1762	440	515	3.16
20	30	220	242	1.1	−35	779	1690	519	459	3.36
21	27.5	210	231	1.1	−25	848	1895	467	386	3.25
22	27.5	200	231	1.155	−30	933	1711	533	357	3.71
23	25	206	228	1.1122	−35	910	1863	501.15	750	3.5
24	25	200	228	1.14	−35	979	1991	541	366	3.84
25	25	205	230	1.122	N/A	926	1846	513	445	3.51
26	25	200	225	1.125	N/A	942	1898	518	397	3.4
27	25	201	222	1.1045	N/A	930	2098	503	496	3.22
28	27.5	201	222	1.1045	N/A	876	1796	488	477	3

Table 6.
Design optimization test matrix.

thickness, yoke plate material and corner radius. This allowed for a systematic testing procedure to evaluate the performance of the different configurations product of the multiple design parameter combinations.

2.6 Section "A": ellipsoidal optimization

Section "A" was found to offer superior structural properties compared to the other designs analyzed in this project. This section allows for the smoothest stress distribution along the composite winding. However, as previously stated, the design exploration aims to find a cross section with a smaller height, such as section "E" (**Figure 5**). Given the poor performance of the custom designed sections (B–E, E1–E4) a new design exploration was carried out by defining the main geometry as an ellipse. By varying the major and minor axis dimensions, subsequently, a more systematic approach was adopted for the optimization of the cross section. Section "A" was used as the starting point, and from there, the design variables (**Figure 2b**) were parameterized. The test matrix for the ellipsoidal optimization of the cross section is presented in **Table 6**. Configuration 13 is the previously optimized circular section and is used as the baseline for this optimization. The objective is to reduce the overall height and material usage.

3. Results

3.1 Section "A" optimization results

As part of the optimization process intended to reduce the overall height and material usage of the pressure vessel's cross section, the press cylinder was removed from the design. A noticeable lack of stress near the yoke plates interface (**Figure 6a**) encouraged the addition of an angled cut (non-horizontal) interfaces between the yoke plates. The advantages of using a non-horizontal yoke plate contact interface are visible in the results for design configurations 2, 8, 9 (**Figure 6b, 9c**, & **9d**, respectively), as the stress near the contact of the yoke plates increases substantially but remains below the material yield point. A decrease in the press cylinder stress was also achieved, so the press cylinder thickness was gradually reduced (**Figur 6c**) until it was completely removed from the assembly. With the press cylinder removed, we concluded that the angled yoke plates' interfaces were not essential, since a simpler horizontal allowed for stress magnitudes within the required limits, as depicted in **Figure 6e** (configuration 13). This resulting cross section design obtained from the optimization process was set as the base for the subsequent optimization procedures, and the finite element model was modified to account for the removal of the press cylinder, as described in 2.3.7 & 2.3.8.

3.2 Section "B," "C," "D," and "E" FEA results

Sections "B," "C," "D," and "E" showed a considerably inferior performance to section "A." However, the surface area of their yoke plates and their overall height (with the clear exception of Section "B") was considerably less than that of section "A." The substandard performance of these sections can be directly associated with the change in curvature of the composite winding due to the change in the height-to-length ratio for the overall cross section. As this ratio increases (height < length), the curvature radius decreases, and thus, stress concentrations appear in the

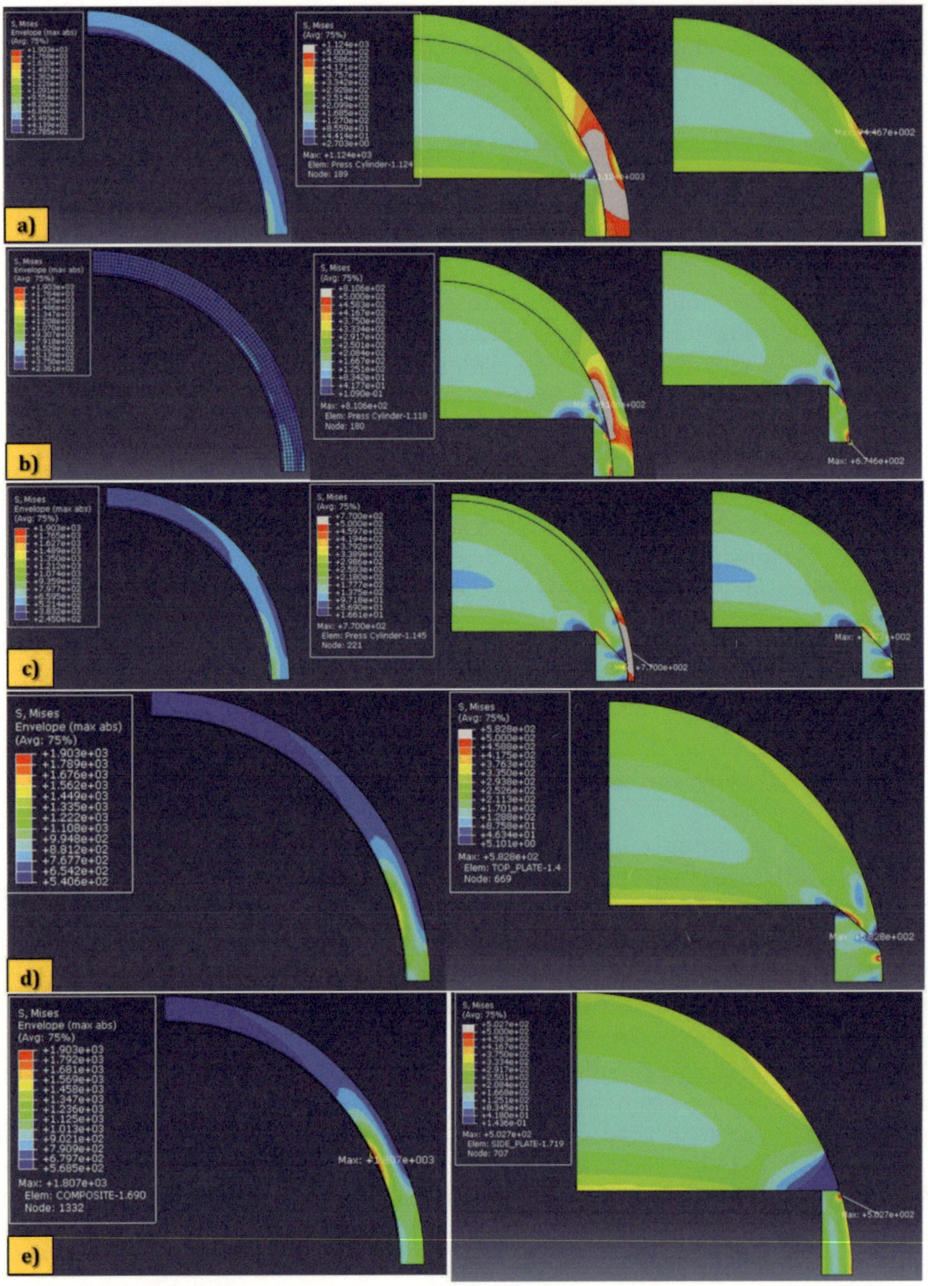

Figure 6.
Stress results for ellipsoidal section design configurations: (a) configuration 1, (b) configuration 2, (c) configuration 8, (d) configuration 9, (e) configuration 13.

composite winding. However, a small length-to-height ratio (height > > length) similar to Section "B" is also undesired, as the stress in the side plates increases considerably.

As is the case with section "A" results, a noticeable lack of stress near the yoke plates interface is observed in the results for sections "B," "C," and "D". Likewise, the stress magnitude on the side yoke plates is less than that of the main yoke plates. A non-horizontal angled cut contact interface could have been implemented for these sections, however, the stress on the composite winding is well above its limit in all sections, and thus, the angled contact interface would've been of little use for these configurations as they were proposed. The results for these sections are presented in **Figure 7**.

3.3 Section "E" variants FEA results

The section E variants appeared to be the ideal candidates for an optimized designed. This was not the case, as the E section proved not to be strong enough to withstand the applied pressure load. The failure mechanism was primarily located in the composite winding, at the inner surface of the corner rounds. To reduce the stress at this location, the corner radii must be increased, although to achieve this, the height of the yoke plates must be increased, which represents a step back

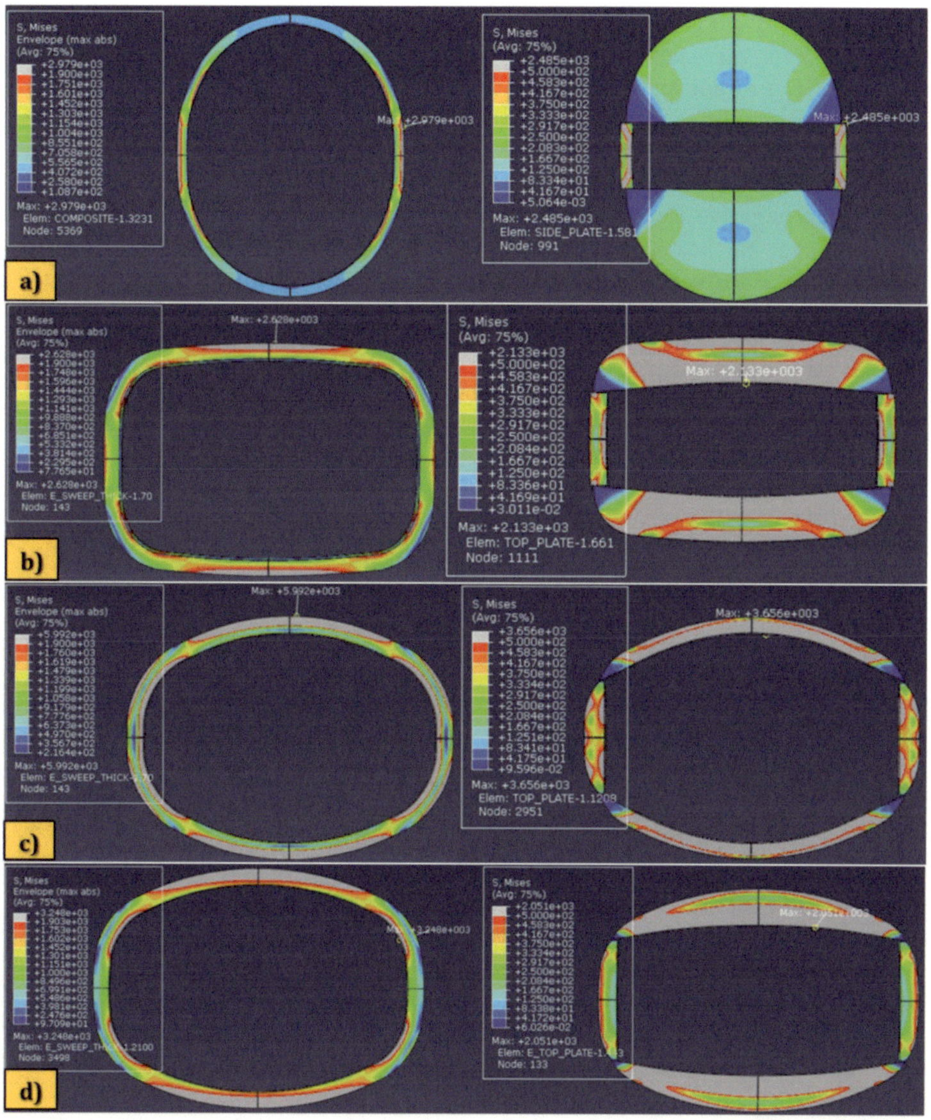

Figure 7.
Stress results for: (a) section "B," (b) section "C," (c) section "D," (d) section "E".

toward the original circular cross section design. We concluded that cross sections that tightly wrap around the inner slot where the pressure is applied are most likely to stress beyond the permissible limit, primarily due to high stress concentration around the sharp radii near the corners. Due to the failure of all the design configurations proposed in the test matrix for Section "E" variations (**Table 5**), only the stress results for the baseline configurations for each custom Section E1–E4 are provided in **Figure 8** to illustrate the stress concentration areas in the different cross section designs.

As stated before, none of the custom design sections (B–E & E1–E4) resulted in feasible designs. Returning to the last feasible configuration (Section "A," design configuration 13), a new design exploration was carried out by defining the main geometry as an ellipse. By varying the major and minor axis dimensions, an ellipsoidal shape with a smaller height was achieved. The design intent and evolution are shown in **Figure 9**, where stress results for key design iteration are provided. The minor axis is reduced considerably during the first configurations, increasing the length-to-height ratio and consequently the stress responses. The minor axis had to be increased back again until the design stress limits were met. Different non-horizontal angled cut contact interfaces were also tested, such as, 45° and −35° interface angles. Based on results and the original idea behind the concept, we concluded that the non-horizontal contact interface must have a negative angle (on the right side of the vertical symmetry plane) with respect to the horizontal plane to be beneficial for the design (achieve load transfer from yoke plate to side yoke plate). The stress results demonstrate that the length-to-height ratio is an important design parameter, which must be close to 1 to achieve good stress distributions along the composite winding. Regardless of the final configuration length-to-height ratio being close to one, i.e., the cross section being almost circular, a considerable reduction in cross section height was achieved.

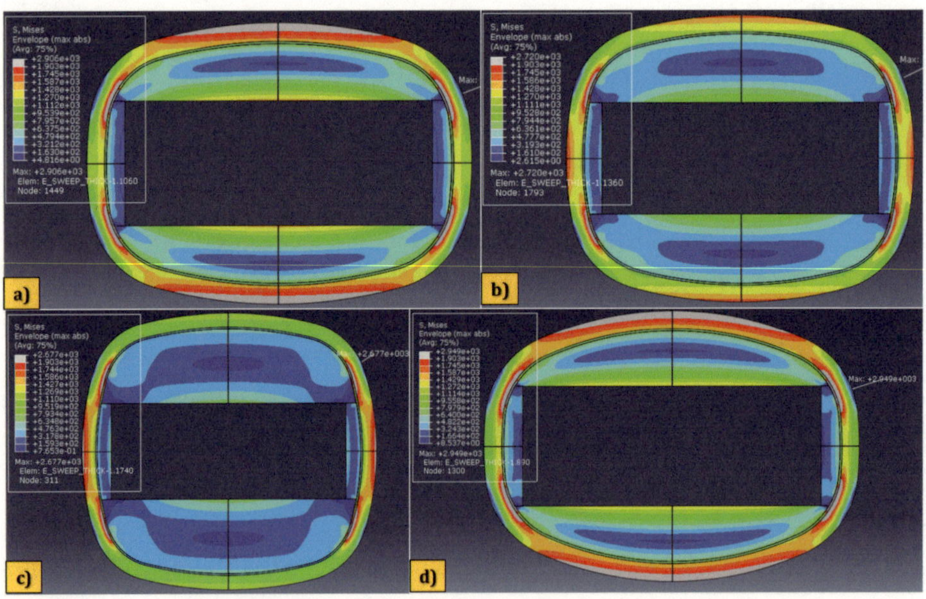

Figure 8.
Stress results for section "E" variations design configurations: (a) section E1, (b) section E2, (c) section E3, (d) section E4.

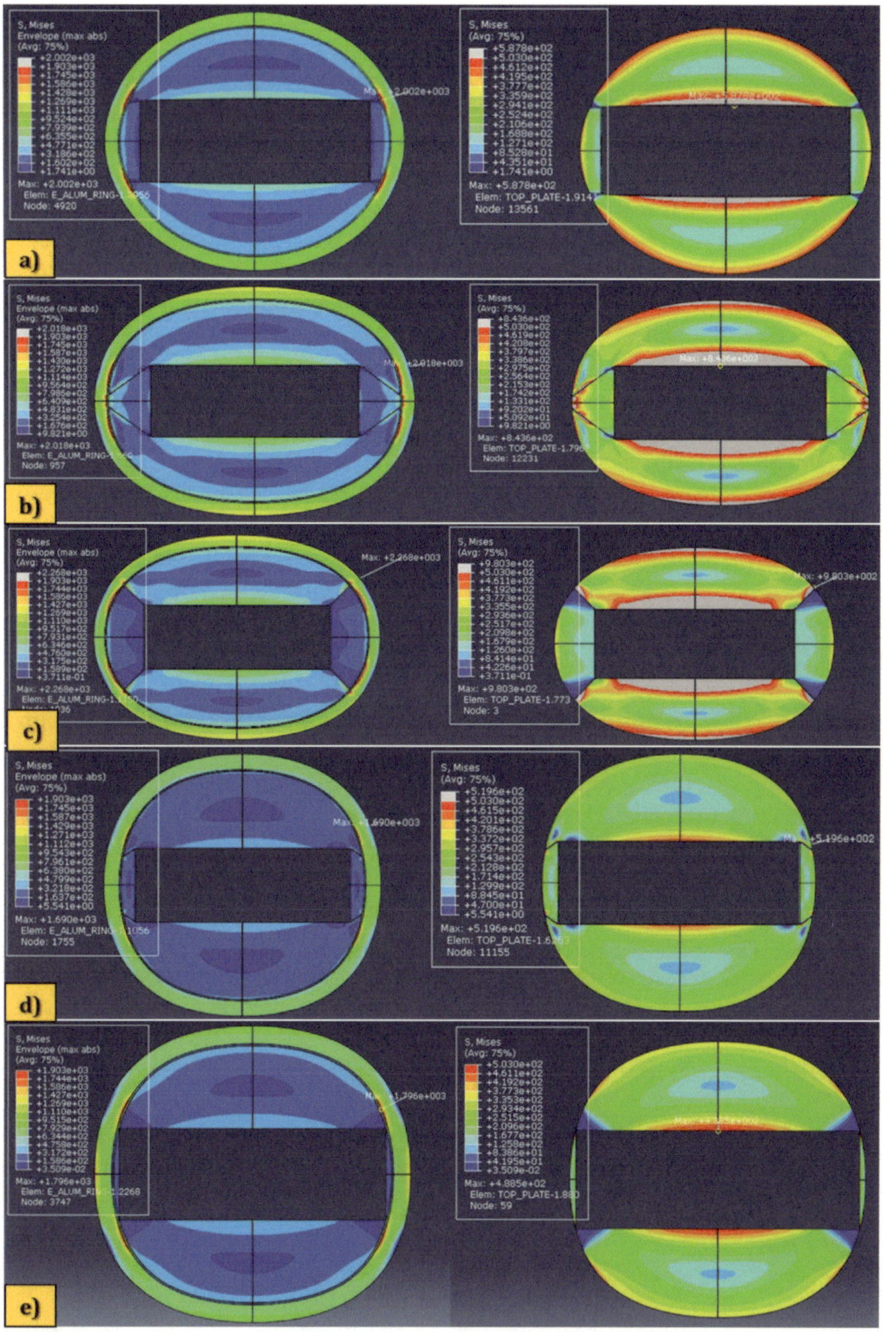

Figure 9.
Stress results for the design configurations produced during the ellipsoidal optimization procedures. (a) Configuration 15, (b) configuration 17, (c) configuration 18, (d) configuration 20, (e) configuration 28.

3.4 Ellipsoidal optimization of circular section "A" FEA results

3.5 Optimization results

A feasible design configuration (**Figure 10**) resulted from this design exploration (marked green in **Table 5**). At first view, it appears to be nearly circular, although it is not. The section has a length-to-height ratio of 1.1045 while maintaining stress concentrations within the allowable limit. The optimized curvatures resulting from the specially selected design parameters provide a reduction of up to 23% (139 mm) on the total cross section height when compared to the original Section "A" design (including press cylinder) (602 mm), or 14% when compared to the optimized circular section "A" without the press cylinder. Similarly, the overall height of the yoke plates was reduced 25% with respect to the original section "A." One particular drawback of the proposed design is the necessary increase in composite winding thickness. A more detailed sensitivity analysis can help determine the optimal composite winding thickness between the bounds set forward by this work in order to adjust the maximum stress on both the composite winding and yoke plates to be equal to the predefined design stress (yield stress).

Note that the yoke plate surface area subjected to pressure is twice as large as that of the side yoke plates. Thus, a greater net force is exerted on the top and bottom yoke plates when compared to the side plates. The applied pressure load produces outward displacement or expansion of the whole structure. However, the composite winding tends to expand more in the vertical direction due to the greater surface area producing a higher total magnitude of force. The vertical displacement generates a contraction of the sides of the composite structure, which opposes the expansive behavior caused by the lateral pressure load. Consequently, the resultant lateral forces acting upon the sides yoke plates have a considerably smaller magnitude than that of the vertical force. This allows for the use of very side yoke plates, as shown in configuration 28. A non-horizontal angled cut is not practical for these design configurations due to the reduced width of the side yoke plates. Otherwise, it may well be possible to further reduce yoke plate height by means of a non-horizontal yoke plate contact interface.

Note that the increase in composite winding thickness may induce additional manufacturing costs, which must be studied in detail. The overall advantage of having a reduced cross section height must be of significant value to justify the increase in cost due to additional composite material. A through manufacturing analysis would be beneficial to determine whether the cost of increasing the composite material usage or increasing the overall yoke plate size yields the best possible outcome.

3.6 ANOVA: analysis of variance of design variables

Finite element analysis results of each parameterized configuration proposed in the test matrix for Section "E" variations were statistically tested to estimate the

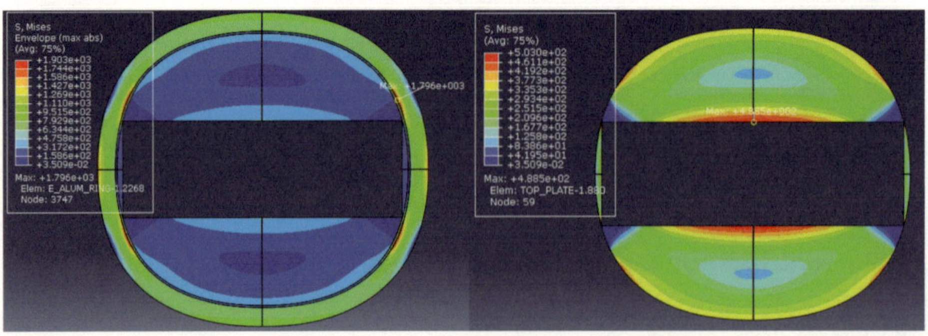

Figure 10.
Ellipsoidal optimization design configuration 28 stress results.

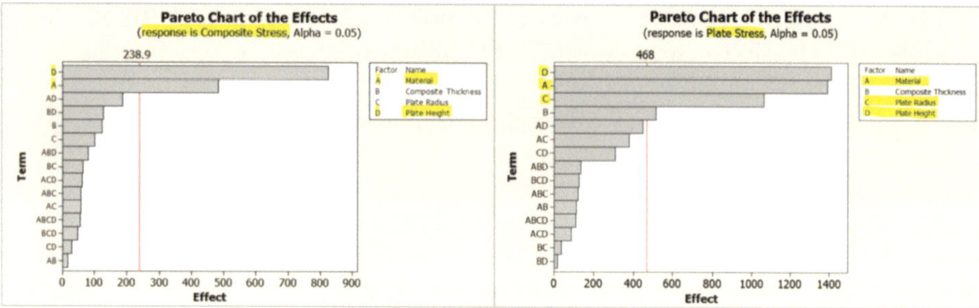

Figure 11.
Effect of design parameter on composite winding stress and yoke plate stress respectively.

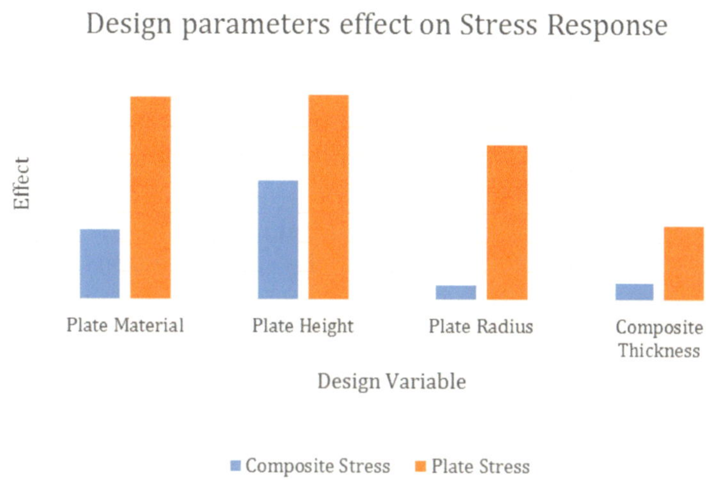

Figure 12.
Accumulative effect of design parameter on composite winding stress and yoke plate stress.

effect of each design variable and their possible combined effects on the maximum stress. An analysis of variance (ANOVA) of the stress response at a specific point in the geometry of each design configuration exposes which design parameter has the greatest effect on the composite winding and yoke plate stresses of the model, respectively (shown in the form of Pareto charts in **Figure 11**). In most cases, plate height has the greatest effect on both stress and displacement responses, followed by the yoke plates material. Aluminum yoke plates show better performance than their steel counterparts.

The accumulated effects in **Figure 11** are plotted in **Figure 12** to illustrate the principal and most sensitive response. It can be concluded that Yoke plate stress is the most sensitive response to changes in the design parameters for custom sections E1–E4.

4. Conclusions

In this study, a multi-physics approach was employed to derive the performance of a type of pressure vessel. Pressure loads were transferred asymmetrically through multiple materials and geometries. The mechanical properties and FEA of various section types were investigated.

The results of the study demonstrate that pressure containment of 165.48 MPa (24,000 psi) is feasible using carbon fiber bobbin wound over a 7075-T6 yoke plate. The simulations show that asymmetric internal loading of a rectangular high-pressure zone containing working fluid does indeed create localized hot spots of pressure in both the containment windings and the associated yoke plates. The primary surprise is that bridge-like bending deflection in the yoke plates is one of the major concerns of the system. This bending deflection caused catastrophic containment winding failure to occur at the centerline of the outer containment windings. Yoke plate deformation would also ultimately cause forming chamber deflection and leakage.

The study reinforces the idea that a near-circular cross section provides the best overall stress distribution throughout the composite winding. However, this comes at the expense of increasing the overall size of the yoke plates. By modifying the design parameters, principally the yoke plates' height, and adjusting the composite winding to seamlessly wrap around the new configuration, a reduction in the yoke plate size is achieved while attaining stress concentrations below each material limit.

L/H ratio defines the overall curvature of the composite, and the closer it gets to 1, the better the performance. However, the stress in the composite is most of the time less than its strength. The main problem has to do with the yoke plates. Early study sections 20–23 have round cross sections, but the stresses induced into the yoke plates exceed the yoke plate's material yield point, thus must be discarded.

Analysis of variance (ANOVA) of the stress responses at a specific point in the geometry reveal that yoke plate section height, yoke plate material, and the overall cross-section curvature radius (length-to-height ratio) are the driving design parameters to achieve a successful and improved solution.

One of the key findings of relevance is the use of an angled wedge integral to the upper yoke plate to create a positive pressure transfer shear plane with the side yoke plates. This positive surface-to-surface contact area allows for lateral expansion compression pressure loads be transferred into the upper yoke plate as surface tension. This additional tension induced into the upper plate helps to stabilize the "bridge bending" tendency of the upper and lower yoke plates. It also reduces the loads transferred by the side yoke plates into the outer fiber windings by confining lateral yoke plate movement expansion.

After testing greatly varied profiles and cross sections, we conclude that a near-round ellipse section successfully accommodates a pressure of 165.48 MPa (24,000 psi). This is using mildly interfacing yoke plates to balance the structure. This solution is conceptually similar to a traditional legacy design section configuration. However, this conclusion is a point of reference - not a final design. It should be considered a proof-of-concept validation point that illustrates that a cross section can be developed to meet the functional performance criteria required using carbon fiber and aluminum in place of steel. As a result, the section as it stands offers a significant reduction in weight over the traditional "steel over steel" fabrication method. A thin-wall section, light weight, reduction or elimination of press cylinder wall achieves a 65% overall weight reduction in similar design due to a shift from steel wound with high tensile steel tape to aluminum wound with carbon fiber.

5. Future perspectives

The pressure containment vessel cross section design can be further optimized by using more organic curve geometry instead of the more traditional arc or ellipse curvatures to define it. Smooth and progressive curve curvatures are expected to allow more gentle and controlled application of pressure load forces.

An improved lower profile design may be possible with additional design revisions. It is believed that a design that has a near vertical yoke plate outer side wall surface to surface interface that also uses the interlocking yoke slip plate design concept developed may produce a better overall performance.

The final functional design will also need to accommodate various voids for fluid ingress and egress as well as valves, electronics and assembly. The use of surface typology optimization will allow the extraction of as much unnecessary sectional mass as possible. This will be helpful in establishing the correct location of these voids.

Author details

Bo C. Jin[1]*, Xiaochen Li[1], Karl Neidert[2] and Michael Ellis[3]

1 Aerospace and Mechanical Engineering, University of Southern California, United States

2 Karl Neidert and Associates, United States

3 Ellis Industrial Design, San Diego, CA, United States

*Address all correspondence to: bochengj@usc.edu

References

[1] Emprotech Corp. Emprotech Corp. [Online]. Available from: http://www.enprotech.com/

[2] SIMULIA Dassault Systems. ABAQUS Theory Manual. Providence; 2012

[3] Jin B, Pelegri A. Three-dimensional numerical simulation of random fiber composites with high aspect ratio and high volume fraction. Journal of Engineering Materials and Technology. 2011;**133**:41014

[4] Wu H, Xu W, Shan D, Jin B. An extended GRN model for low stress triaxiality and application in spinning forming. Journal of Materials Processing Technology. 2019;**263**:112-128

[5] Jain A, Jin B, Nutt S. Mean field homogenization methods for strand composites. Journal of Composites Part B. 2017;**124**:31-39

[6] Jin B, Li X, Jain A, Wu M, Mier R, Herraez M, et al. Prediction of stiffness of reused , carbon fiber/epoxy composite oriented strand board using finite element methods. Recycling materials & structures. Feature article. Society for the Advancement of Material and Process Engineering (SAMPE) Journal. 2017;**53**:32-39

[7] Jin B, Li X, Jain A, Herraez M, Gonzalez C, LLorca J, et al. Development of a finite element model for reused carbon fiber epoxy composite oriented strand board. In: Proceedings of SAMPE 2016; Long Beach. 2016

[8] Jin B, Li X, Mier R, Pun A, Joshi S, Nutt S. Parametric modeling, higher order FEA and experimental investigation of hat-stiffened composite panels. Journal of Composite Structures. 2015;**128**:207-220

[9] Jin B, Joshi S, Pun A, Nutt S. Design sensitivity of hat-stiffened composite panels. In: Proceedings of American Society for Composites Conference 2014; San Diego. 2014

[10] Jin B, Liu W, Patel H, Nutt S. Application of MSC NASTRAN UDS in modeling and analysis of a hybrid composite reinforced conductor core. In: Proceedings of MSC NASTRAN Conference 2013; Los Angeles. 2013

[11] McLaughlan PB, Forth SC. Composite Overwrapped Pressure Vessels, A Primer. Houston; 2011

[12] Ma H, Xu W, Jin B, Shan D, Nutt S. Damage evaluation in tube spinnability test with ductile fracture criteria. International Journal of Mechanical Sciences. 2015;**100**:99-111

[13] Wu H, Fan G, Jin B, Geng L, Cui X, Huang M. Fabrication and mechanical properties of TiBw/Ti-Ti (Al) laminated composites. Journal of Materials and Design. 2016;**89**:697-702

[14] Wu H, Fan G, Jin B, Geng L, Cui X, Huang M, et al. Enhanced fracture toughness of TiBw/Ti3Al composites with a layered reinforcement distribution. Journal of Materials Science and Engineering: A. 2016;**670**:233-239

[15] Wu H, Jin B, Geng L, Fan G, Cui X, Huang M, et al. Ductile-phase toughening in TiBw/Ti-Ti3Al metallic-intermetallic laminate composites. Journal of Metallurgical and Materials Transactions A. 2015;**46**:3803-3807

[16] Chen Z, Huang T, Jin B, Hu J, Lu H, Nutt S. High yield synthesis of single-layer graphene microsheets with dimensional control. Journal of Carbon. 2014;**68**:167-174

[17] Huang T, Li T, Xin Y, Jin B, Chen Z, Su C, et al. Preparation and utility of a self-lubricating and anti-wear graphene oxide nano-polytetrafluoroethylene hybrid. Journal of RSC Advances. 2014;**4**:19814

[18] Huang T, Jin B, Li X, Li T, Nutt S. Experimental and finite element analysis study of load carrying capacity of modified polyimide. In: Proceedings of SAMPE 2013; Wichita. 2013

[19] NASA. A Theoretical Investigation of Composite Overwrapped Pressure Vessel (COPV) Mechanics Applied to NASA Full Scale Tests. Houston; 2009

[20] Jin L, Jin B, Kar N, Nutt S, Sun B, Gu B. Tension-tension fatigue behavior of layer-to-layer 3D angle-interlock woven composites. Journal of Materials Chemistry and Physics. 2013;**140**:183-190

[21] Jin L, Niu Z, Jin B, Sun B, Gu B. Comparisons of static bending and fatigue damage between 3D angle-interlock and 3D orthogonal woven composites. Journal of Reinforced Plastics and Composites. 2012;**31**(14):935-945

[22] Ma Q, Wang K, Wang S, Liu H, Jin B, Jin L, Ma P. Tensile damage mechanism of 3-D angle-interlock woven composite using acoustic emission events monitoring. Autex Research Journal. 2018;**18**(1):46-50

[23] Jin B, Li X, Jain A, Carlos G, LLorca J, Nutt S. Optimization of microstructure and mechanical properties of composite oriented strand board from reused prepreg. Journal of Composite Structures. 2017;**174**:389-398

[24] Wu M, Jin B, Nutt S. Processing and performance of recyclable composites for wind turbine blades. Advanced Manufacturing: Polymer & Composites Science. 2019;**12**:12-28

[25] Jin B, Li X, Jain A, Wu M, Mier R, Herraez M, et al. Mechanical properties and finite element analysis of reused UD carbon fiber/epoxy OoA VBO composite oriented strand board. In: Proceedings of SAMPE 2016; Long Beach. 2016

[26] Jin B, Li X, Wu M, Jain A, Jormescu A, Gonzalez C, et al. Nondestructive testing and evaluation of conventional and reused carbon fiber epoxy composites using ultrasonic and stitched micro-CT. In: Proceedings of SAMPE 2016; Long Beach. 2016

[27] Jain A, Jin B, Li X, Nutt S. Stiffness predictions of random chip composites by combining finite element calculations with inclusion based models. In: Proceedings of SAMPE 2016; Long Beach. 2016

[28] Jain A, Jin B, Nutt S. Development of chip composites with improved mechanical performance attributes using hybrid multi-scale modelling methods. In: Proceedings of CAMX 2016; Anaheim. 2016

[29] Jin B. Recent development of reused carbon fiber reinforced composite oriented strand boards. Chapter 8 of Book: Recent Developments in the Field of Carbon Fibers. IntechOpen. DOI: 10.5772/intechopen.71346

[30] Zoltek. PANEX 35 commercial carbon Fiber. In: Material Specs Sheet.

Section 2

Combustion, Air and Noise Pollution

Chapter 3

The Role of CFD in Modern Jet Engine Combustor Design

Zhi X. Chen, Ivan Langella and Nedunchezhian Swaminathan

Abstract

Recent advances in the application of computational fluid dynamics (CFD) for turbulent combustion with the relevance for gas turbine jet engines are discussed. Large eddy simulation (LES) has emerged as a powerful approach to handle the highly turbulent, unsteady and thermochemically non-linear flows in the practical combustors, and it is a matter of time for the industry to replace the conventional Reynolds averaged Navier-Stokes (RANS) approach by LES as the main CFD tool for combustor research and development. Since combustion is a subgrid scale phenomenon in LES, appropriate modelling is required to describe the SGS combustion effects on the resolved scales. Among the various available models, the flamelet approach is seen to be a promising candidate for practical application because of its computational efficiency, robustness and accuracy. A revised flamelet formulation, FlaRe, is introduced to outline the general LES methodology for combustion modelling and then used for a range of test cases to demonstrate its capabilities for both laboratory burners and practical engine combustors. The LES results generally compare well with the experimental measurements showing that the important physical processes are captured in the simulations.

Keywords: jet engine combustion, computational fluid dynamics, large eddy simulation, turbulent combustion modelling, FlaRe

1. Introduction

Over the past half a century, the combustor design for jet engines has been driven to meet the increasingly higher standards of thermal efficiency, emission reduction and power-to-size/weight ratio. Consequently, the operating conditions have experienced a dramatic change, e.g., the operating pressure increasing from several bars to few tens of bars, combustor inlet temperature from about 500 to nearly 1000 K, and turbine inlet temperature from just above 1000 to almost 2000 K in today's turbo-fan engines [1]. Despite the severe change in the fluid-mechanical and thermodynamic properties of the combustor internal flow, the combustor length and frontal area are still strictly limited by the design factors for other engine components. Also, the current unprecedented demand for global travels requires a much longer lifespan for aero engines, typically many tens of thousands of hours without major maintenance. To meet these requirements, jet engine manufacturers continuously seek avenues for a reliable, efficient and economical combustor design cycle.

The emergence of computational fluid dynamics (CFD) has made computer-aided design an integral part of the gas turbine (GT) combustor design process [2].

Compared to the expensive experimental tests, which provide only global information (e.g., stability, outlet properties), CFD is much cheaper to run and, most importantly it can be repeated during the design process to examine the effects of small design changes. Thus, it is attractive for practical applications. Over the recent years, there has been a significant increase in the investment from the industry for the development of CFD tools, but the challenges remain because fully resolving the turbulent reacting flows in practical jet engines using direct numerical simulation (DNS) is still far beyond our reach. The Reynolds averaged Navier-Stokes (RANS) approach has been broadly used as the main CFD tool for practical combustor design in the last few decades. Because all the scales are modelled in RANS, only mean flow quantities are computed leading to a cheap and fast calculation. For steady combusting flows, good accuracy can be achieved if the correlation between the fluctuating quantities is handled correctly. The drawback for RANS is also obvious since the transient phenomena, such as ignition, flashback, thermoacoustics and blow-off, cannot be captured by the mean flow calculation. While these phenomena are of high interest for the industry, there is an increasing need for predictive CFD tools. Large eddy simulation (LES) is recognised as a promising candidate as the energy-containing flow structures are directly resolved by the numerical grid and the subgrid scales (SGS) are modelled. In general, with respect to RANS the main advantages of LES are twofold: the capability to capture the transient phenomena and a better prediction of mixing. Thus, the LES modelling paradigm is of interest here.

The level of resolved scales in LES dictates the computational cost, i.e., the more resolved the more expensive; however, it only partly determines the overall accuracy. A significant part of the accuracy is attributed to SGS modelling, which represents the influence of subgrid motion on the resolved scales. For the velocity field, this influence usually appears through the SGS eddy viscosity and in a well-resolved LES, i.e., typically over 80% of the turbulent kinetic energy is resolved [3], the SGS effect is relatively small. For the flame, however, chemical reactions occur at scales smaller than the typical LES filter size and thus are SGS phenomena requiring closure models. The major challenge is how to model the SGS interaction between turbulence and chemistry, with the latter involving a large number of species and reactions (typically many thousands for common jet fuels). Consequently, this makes it practically unfeasible to directly integrate detailed chemical kinetics into the LES. Finding a computationally efficient model with good accuracy and robustness for the SGS turbulence-chemistry interaction has been a central topic for turbulent combustion research in the last two decades, and a number of approaches are available in the literature. Extensive review of these approaches is beyond the scope of this chapter as detailed reviews are available elsewhere, see for example [2–5], and the focus here is to showcase the current capabilities of combustion LES modelling for practical applications.

The rest of this chapter is organised as follows. Section 2 describes the modelling challenges in LES of gas turbine combustion and a representative approach to tackle these challenges. The validation test cases for this approach are presented in Sections 3 and 4 for laboratory and practical burners respectively. The conclusions are summarised with a future outlook in Section 5.

2. State-of-the-art LES modelling for gas turbine combustion

With the advent of high performing computing, large eddy simulation has become increasingly popular to investigate complex and unsteady physics in gas turbines due to its versatility in capturing time-dependent phenomena and in controlling the computational effort at the same time. This is achieved by varying the LES filter, whose shape is implicit and depends on local mesh size, SGS model and numerical scheme. It is generally accepted that the filter size is proportional to the local cell volume, $\Delta \approx \mathcal{V}^{1/3}$. The larger is Δ, the stronger is the contribution of the SGS modelling to the results, with the generally accepted rule that at least 80% of the turbulent kinetic energy should be resolved [3]. The general LES equation for a quantity φ takes the form, in Einstein's notation,

$$\frac{\partial \overline{\rho}\tilde{\varphi}}{\partial t} + \frac{\partial \overline{\rho}\tilde{u}_i\tilde{\varphi}}{\partial x_i} = -\frac{\partial \overline{p}}{\partial x_j} + \overline{\mathcal{D}} + \frac{\partial \overline{\tau}}{\partial x_i} + \overline{S}, \tag{1}$$

where the overbar denotes a spacial filtering operation. For density varying flows the filtering operation would lead to additional unclosed terms and to avoid this a density-weighted, or Favre-filtered operator is introduced, defined as $\tilde{\varphi} = \overline{\rho\varphi}/\overline{\rho}$. Note that as $\overline{\varphi} \neq \tilde{\varphi}$, one has to be mindful when comparing CFD results with measurements in regions of strong density gradients, and this will be discussed again in Section 4. The terms on the LHS of Eq. (1) represent time variation and convection of the filtered quantity $\tilde{\varphi}$. The pressure derivative on the RHS in present only if φ is a velocity component, u_j. The application of the LES filter to the convective term leads to the appearance of the term $\overline{\tau} = \overline{\rho}\left(\widetilde{u_i\varphi} - \tilde{u}_i\tilde{\varphi}\right)$ representing the effect of subgrid processes, and this term requires modelling. Well accepted models are available, e.g., the Smagorinsky model [6] for SGS stresses and the gradient transport model for unresolved scalar fluxes [2]. The filtered diffusion term, $\overline{\mathcal{D}}$, takes the form of a Laplacian and also may need modelling. However, this modelling is generally irrelevant for most high-turbulent conditions proper of gas turbines since the SGS turbulent diffusion is dominant. The last term in Eq. (1) is the filtered source term representing compressibility, or gravitational, or evaporation or heat release effects. For the equation of species the source term is the reaction rate and its modelling is the objective of this section.

For industrial gas turbine conditions the flame thickness, δ, is generally small, so to keep the computational cost affordable for industrial operations, Δ is always larger than δ and thus the combustion processes are entirely at SGS level. Note that it is not generally a problem to satisfy the 80% rule, since the evaluation of the turbulent kinetic energy excludes the combustion dilatation effects [3, 7]. In light of the above considerations, recent development of combustion modelling has gone in two directions. One is to include the full thermochemistry into the modelling, and at the expenses of computational cost. These types of modelling are usually unaffordable for industrial design purposes, but together with DNS methodology they can be used to investigate complex phenomena and SGS processes in laboratory scale burners. The other approach is more industrial-oriented, where the objective is to keep the computational cost to a minimum so that the model is usable for practical combustors. The thermochemistry is included through statistical or geometrical means. Based on this distinction, these two directions can be categorised respectively as non-flamelet and flamelet approaches. The gap in the accuracy between these two categories has reduced with time and recent advances have shown that flamelet-based models are capable of representing the complex flow features in gas turbines despite the limitations of their underlying assumptions. These models are reviewed in a number of works, see for example [2, 4, 8]. Because of the relevance for gas turbine applications, only the flamelet category is discussed here. Within the flamelet category there are geometrical and statistical approaches. Although in both cases the thermochemistry is computed *a priori*, the assumptions behind geometrical models usually lead to the need of relatively large mesh in order to achieve a significant increase of accuracy, see for example the discussion on thickened flame [9, 10] model in Section 4, or the introduction of additional complexity to smooth the G-equation in level set models [11–13]. Thus, the use of this category of models for industrial applications is still unclear. The statistical models within the flamelet approach include the turbulence-chemistry interaction using probability density functions (PDFs), which are typically presumed and thus they do not incur additional computational effort. Although additional equations are still necessary depending on the particular model, these are generally computationally cheap to solve because the filter size can be kept larger than the flame thickness, at least in principle, as long as the presumed PDF used is able to represent the statistical behaviour at scales smaller than Δ correctly. The potential of flamelet-based models for GT applications has thus to pass through a deep understanding of the SGS processes, which has been the focus of research in the last 30 years. Only in recent years, however, revised flamelets formulations for LES have demonstrated potentials to bridge the gap that separated them from models directly accounting for the thermochemistry.

In flamelet models, the turbulent flame is seen as an ensemble of thin, one-dimensional structures (flamelets) which are wrinkled by turbulence; turbulent eddies are either not small enough to penetrate into the flame and alter its internal structure, or they do not last long enough. Therefore, the thermochemistry can be computed *a priori* through one-dimensional computations and then parameterised using a set of control variables. For partially premixed combustion these are usually a variable to track the rate of mixing, and another to track the reaction progress. Other parameters can be introduced to include additional effects such as pressure variation, non-adiabaticity, strain, etc. The 1D laminar flames can be of any type, e.g. they can be premixed or diffusion flame. Premixed flamelets are, however, more versatile for partially premixed combustion as strain, reaction progress and mixing can be controlled independently; thus they can, in principle, be used for situations involving local extinctions. The big challenge for using flamelet models for GT combustion conditions is that turbulence is extremely high and the smallest eddies can penetrate and affect the internal flame structure, thus invalidating the flamelet hypothesis. This can happen when the Karlovitz number, $Ka = \tau_c/\tau_\eta \gg 1$, where τ_c and τ_η are chemical and the Kolmogorov time scales respectively. Nevertheless, a number of relatively recent works (see for example [14–16]) have shown that flamelet structures are present at GT combustion conditions, but distributed over a wider region yielding a thicker flame brush rather than a thicker flame.[1] This is because small eddies may not have enough energy to impart significant changes to the flame structures [17, 18] and thus the limits of applicability of premixed flamelets are unclear [2, 19].

For stable GT combustion conditions the pressure across the flame does not vary significantly and thus different flamelets for different pressures are not typically computed. The effect of heat losses is also generally taken to be small for combustion modelling purposes. The effect of strain on a premixed flamelet is well accepted to be important in the case of RANS simulations, however its relevance for LES is more controversial. Recent findings [20, 21] show that, since part of the strain is resolved in the LES, its effect on the flame is implicitly captured as long as the local mesh size is appropriate. These preliminary findings have been confirmed by GT calculations [22–24] and show that strained flamelets are unnecessary at least for the conditions considered. For an industrial perspective where the reduction of computational cost is essential, these recent findings open the way to effective use

---

[1] The flame brush is the time-averaged high temperature region. Hence, it can be thick despite the flame being thin when the flame moves or is distributed spatially.

of flamelets. Following the above discussion, the parameterisation of a model based on unstrained premixed flamelets at various equivalence ratios reduces to only two controlling variables: mixture fraction, ξ, to track the amount of mixing (thus the equivalence ratio) and a progress variable, c, to track the reaction progress. The first is usually defined using Bilger's expression [25], and assuming that all species have the same mass diffusivity, its transport equation is:

$$\overline{\rho}\frac{D\tilde{\xi}}{Dt} = \nabla \cdot \left(\overline{\rho}\mathcal{D}_{eff}\nabla\tilde{\xi}\right) + \overline{\dot{\omega}}_S, \tag{2}$$

where $\mathcal{D}_{eff}$ is the effective molecular diffusivity (sum of filtered diffusivity and the SGS contribution due to the filtering of the non-linear terms). The source term, $\overline{\dot{\omega}}_S$, is for the evaporation of fuel droplets [23, 26, 27]. The progress variable is usually defined as a combination of reactant or product species and varies monotonically from 0 in the reactants to 1 in the products when it is normalised appropriately, although unscaled formulations are often used. A good choice for lean combustion is to define the progress variable as a linear combination of CO_2 and CO [28] mass fractions (normalised by their burnt value). The transport equation for the filtered progress variable is:

$$\overline{\rho}\frac{D\tilde{c}}{Dt} = \nabla \cdot \left(\overline{\rho}D_{eff}\nabla\tilde{c}\right) + \overline{\dot{\omega}}_c^*, \tag{3}$$

where $\overline{\dot{\omega}}_c^*$ is the modified filtered reaction rate, which is tabulated and thus is accessed using the controlling variables themselves during the simulation runtime. This term is expressed as $\overline{\dot{\omega}}_c^* = \overline{\dot{\omega}}_c + \overline{\dot{\omega}}_{np} + \overline{\dot{\omega}}_{ct}$, where $\overline{\dot{\omega}}_c$ is the premixed flame contribution (including mixture stratification). The additional terms represent non-premixed mode contribution and mixed mode due to the interaction of ξ and c gradients, and they appear only for normalised definitions of the progress variable, see additional details for example in [22, 25, 29]. At this point the set of equations would be closed if the above reaction rates depend only on the two controlling variables, as long as a thermodynamic model is provided. However, in this form the effect of wrinkling of the flame by turbulence at the subgrid level is not accounted. This effect is introduced in a statistical way using a presumed subgrid PDF[2]:

$$\overline{\dot{\omega}}_c = \int_0^1\int_0^1 \dot{\omega}(\zeta,\eta)P(\zeta,\eta)d\zeta d\eta, \tag{4}$$

where $\dot{\omega}$ is the laminar reaction rate from the flamelet and ζ and η are the sample space variables for c and ξ respectively. This type of closure was first introduced by Bradley for RANS and non-premixed combustion [30]. The SGS joint PDF can be written as the product of two PDFs as $P(\zeta,\eta) = P(\zeta)P(\eta|\zeta)$. There are various possible choices for the shape of these two SGS PDFs, each with its own advantages and disadvantages, the most common being the Beta PDF and laminar flamelet PDF [31], and it is commonly accepted that these shapes need to be dependent at least on first and second moments. However, for cases involving large turbulence and filter sizes larger than the flame thickness, the Beta PDF was shown to be more

[2] Note that the term PDF in this case does not strictly relates to a probability density function in a statistical sense, since in the LES this operator is used to account for events in space at one time. The use of the term subgrid PDF is thus made for simplicity and analogy to the statistical PDF operator; other authors prefer the term 'filtered density function' to make this distinction.

appropriate in several works (see for example [20, 32–36]). The beta function requires the first and second moments, thus the SGS joint PDF is expressed as $P(\zeta,\eta) = \beta(\zeta;\tilde{c},\sigma_c^2)\beta\left(\eta|\zeta;\tilde{\xi},\sigma_\xi^2\right)$, where σ_c^2 and σ_ξ^2 are the SGS variances of the progress variable and mixture fraction respectively. Note that here the two PDFs are treated independently, which is usually acceptable in LES with an appropriate grid size [37]. The SGS variances obtained using their transport equations are better than using algebraic expressions since convective and diffusive processes are important at subgrid scales [20]. These equations are written as:

$$\overline{\rho}\frac{D\sigma_\xi^2}{Dt} = \nabla\cdot\left(\mathcal{D}_{eff}\nabla\sigma_\xi^2\right) - 2\overline{\rho}\tilde{\varepsilon}_\xi + 2\overline{\rho}\frac{\nu_t}{Sc_t}\left(\nabla\tilde{\xi}\cdot\nabla\tilde{\xi}\right) + \overline{\rho}S, \tag{5}$$

for the mixture fraction variance, and

$$\overline{\rho}\frac{D\sigma_c^2}{Dt} \approx \nabla\cdot\left(\mathcal{D}_{eff}\nabla\sigma_c^2\right) - 2\overline{\rho}\tilde{\varepsilon}_c + 2\overline{\rho}\frac{\nu_t}{Sc_t}\left(\nabla\tilde{c}\cdot\nabla\tilde{c}\right) + 2\left(\overline{c\dot{\omega}} - \tilde{c}\,\overline{\dot{\omega}}\right) \tag{6}$$

for the SGS variance of progress variable. From left to right the various terms in the above equations represent total derivative, effective diffusion, scalar dissipation, turbulent production and source term. The evaporation of droplets contributes to σ_ξ^2 and the subgrid reaction processes contribute to σ_c^2. The latter is closed with an expression consistent with Eq. (4). The evaporation and the spray model in general are out of the scope of this section. For the specific simulations to be presented in Section 4, a Lagrangian approach is used for the two-phase flow, with parcel sampled using a Rosin-Rammler distribution and the Sattelmayer correlation for the initial Sauter mean diameter (SMD) [38]. Only secondary breakup is considered using the process described in [39], and a rapid mixing formulation for the droplet evaporation. More details can be found in [23, 27, 40].

The scalar dissipation rate (SDR) terms in Eqs. (5) and (6) need closure. The linear relaxation model, $\tilde{\varepsilon}_\xi = C_\xi\left(\nu_t/\Delta^2\right)\sigma_\xi^2$, with $C_\xi \approx 2$, is well accepted [5]. Recent works have suggested that this constant is to be revised in case of liquid fuel due to the evaporation source term in Eq. (5) [27]. For the progress variable SGS variance, it is shown in [20] that the reactive term in this equation is of leading order at least for Δ of sizes comparable or larger than the flame thickness, and thus the SDR has to balance the sources coming from reaction and turbulence. Hence, the linear relaxation model is unsuitable on physical grounds. To justify the use of linear relaxation model, a delta or three-delta function is used sometimes instead of the Beta function in Eq. (4) so that $\left(\overline{c\dot{\omega}} - \tilde{c}\,\overline{\dot{\omega}}\right) = 0$ in Eq. (6). This, however, creates a conflict since the meaning of σ_c^2 changes without the reactive term. More recently, revision of the flamelet modelling in the context of LES to take into account the correct reaction-turbulence-diffusion balance led to appearance of more sophisticated, yet simple, model for the SDR of progress variable. One recent development in this sense is the SDR closure of Dunstan et al. [41], which approaches the linear relaxation concept in the limiting behaviour of non-reactive mixture and is thus more recommended. This model has been used in many past studies for different combustion regimes [20, 22, 24, 36] (see also Sections 3 and 4).

The set of equations shown above is used in conjunction with the LES equations for mass and momentum, which are the same for reacting and non-reacting flows, see for example [2, 3] for a more detailed explanation. In principle, the temperature and density fields can be also computed *a priori* using an equation consistent to (4) and accessed in runtime using the controlling variables. In order to account for possible non-adiabatic effects, an energy equation is often solved. For low-Mach conditions it is convenient to use a total specific enthalpy (sum of formation and sensible enthalpies), which is a conserved quantity. The equation for the filtered total enthalpy, $\tilde{h}$, has the same form of Eq. (2) (except for thermal diffusivity in place of mass diffusivity and no source term). The temperature field is obtained using $\tilde{h}$ by inverting the following expression:

$$\tilde{h} = \widetilde{\Delta h_f^0} + \int_{T_0}^{T} \tilde{C}_p(T)\ dT \quad (7)$$

where Δh_f^0 is the enthalpy of formation of the mixture at temperature T_0 and C_p is the heat capacity at constant pressure. This inversion can be performed numerically in different ways, see for example [20, 23, 42], and requires the integration of enthalpy of formation and heat capacity using an equation consistent to Eq. (4). The density is computed via the state equation, where the pressure is often assumed to be constant in low-Mach formulations for numerical stability, except for cases where compressibility effects are important, e.g. thermoacoustic instabilities. The above equations describe the general flamelet formulation with specific details for the FlaRe approach. In the next sections specific test cases relevant for gas turbine engines will be discussed using both laboratory and practical flames, and the FlaRe approach is compared to other combustion model results where they are available.

3. Laboratory burners

For practical jet engines, it is technically difficult and very costly to conduct measurements inside the combustion chamber due to the extremely hostile conditions and complex geometry. Therefore, laboratory burners not only play a crucial role in experimental combustion research but also serve as a main source for CFD model validation. In the past, the majority of the modelling efforts were devoted to flames in simple geometry such as jet flames and bluff-body or swirl stabilised flames in an open environment. Many of these flames have been benchmarked as standard model validation cases documented in the well-known TNF Workshop [43]. Over the last few decades, a large number of combustion models including most of those discussed earlier in Section 2 have been tested using the TNF benchmark flames [44]. Despite the different models used, the computational results converge to a similar level of very satisfactory accuracy when compared with measurements for the main flow and flame statistics. For example, the transient ignition of a lifted methane-air jet flame [45] was simulated using FlaRe [46], conditional moment closure (CMC) [47], thickened flame (TF) [48] and transported PDF with Eulerian stochastic fields (TPDF/ESF) [49] approaches, all showing comparably good agreement with the measurements for the flame upstream propagation. However, this level of general agreement among different models is yet to be achieved for more complex engine-relevant geometry and conditions. Therefore, this section focuses on the state-of-the-art laboratory gas turbine model combustors (GTMCs). In order to demonstrate the current CFD capabilities of tackling the various issues in these combustors, two cases, for single and multiple burner configurations, are considered.

3.1 Single burner with dual swirlers

The dual-swirl GTMC experimentally investigated by Meier et al. [50, 51] at the German Aerospace Center (DLR) is of interest. The schematic of this GTMC

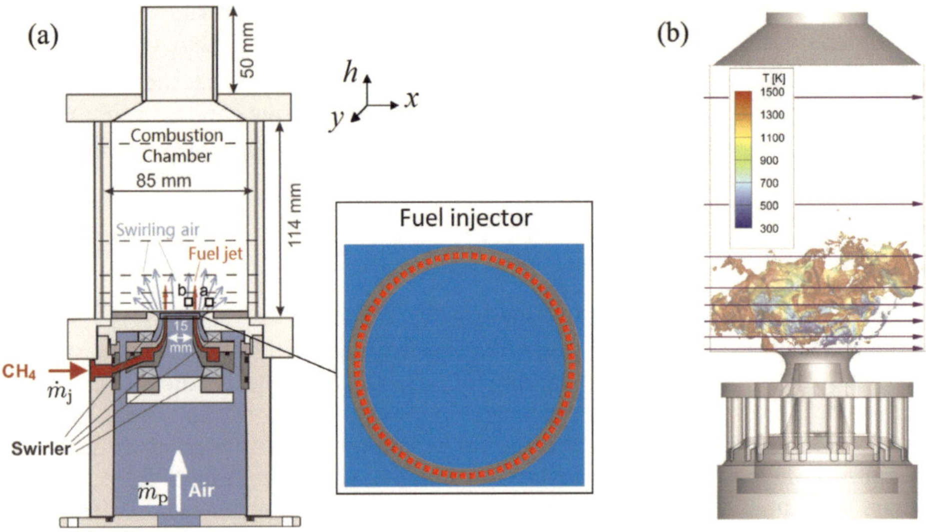

Figure 1.
The DLR dual-swirl combustor: (a) schematic of the experimental setup and (b) typical flame surface marked using $\overline{\dot{\omega}}_c^* = 200$ *kg/m³/s, coloured by temperature.*

[50, 51] and a typical computed flame surface using LES [24] are shown in **Figure 1**. This burner has a single nozzle head with dual-swirl air passages, which is a common design in modern jet engines to achieve fast fuel-air mixing. The methane gas injector was modified from a practical air-blasting liquid fuel injector mounted on the wall in between the two air nozzles. The experiments were conducted at atmospheric pressure and the operating range investigated was from 5 to 35 kW with the equivalence ratio varying from 0.5 to 1.2, which are typical jet-engine relevant conditions [1].

Within the range of conditions operated, a variety of phenomena were observed including flame-vortex interaction [52], self-excited thermo-acoustic oscillations [53] and lean blow-out (LBO) dynamics [54] in the experiments. Three cases detailed in **Table 1** were chosen for experimental study: a thermo-acoustically stable flame, designated as flame A, an unstable flame showing self-excited thermoacoustic oscillations, called flame B, and flame C exhibiting periodic blowout and re-ignition. To investigate the rich physics exhibited in these flames, state-of-the-art laser diagnostic techniques including stereoscopic particle image velocimetry (stereo-PIV), Raman spectroscopy, laser Doppler velocimetry (LDV), OH^*/CH^* chemiluminescence and $OH/CH/CH_2O$ planar laser-induced fluorescence (PLIF), were performed and highly repetitive results were obtained. Hence, these measurements constitute a comprehensive database for rigorous combustion model assessment.

As noted earlier, only typical results are presented here for a brief demonstration of the model performance. **Figure 2** compares the measured and computed mean

Flame	ϕ_{glob}	Z_{glob}	$\dot{m}_p$ [g/s]	$\dot{m}_j$ [g/s]	Swirl number	P_{th} [kW]
A (stable)	0.65	0.037	18.25	0.697	0.9	34.9
B (unstable)	0.75	0.042	4.68	0.205	0.55	10.3
C (approaching LBO)	0.55	0.031	4.68	0.15	0.55	7.6

Table 1.
Summary of operating conditions.

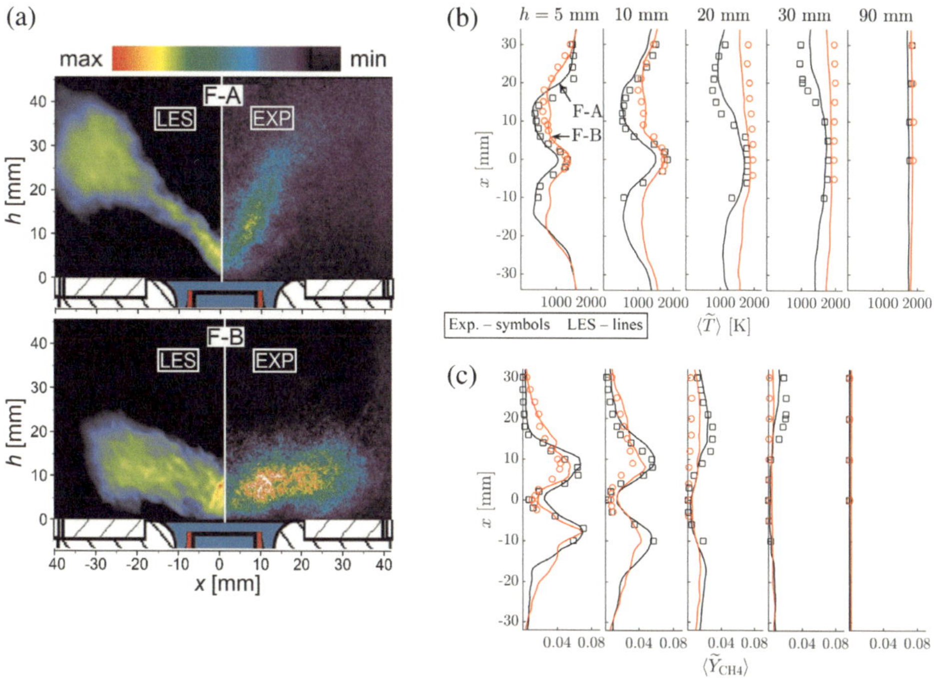

Figure 2.
Typical comparison between LES results and measurements for flames A and B: (a) mid-plane mean concentration of CH radials, mean radial variation of (b) temperature and (c) fuel mass fraction.

reaction zone shape (represented by CH radials), temperature and fuel concentration distributions in the combustion chamber. It can be seen that the overall behaviours of these quantities are captured quantitatively in the simulation, suggesting the accuracy of these CFD calculations has reached a sufficient level for practical design purposes. Here it is of particular practical importance that the change in the flame shape from a V-form in flame A to a flat shape in flame B is correctly reproduced by the LES because the location and distribution of the flame dictates many design factors such as combustor cooling and pollutant emission control, etc. The underlying physical mechanism for this flame shape change involves a fine interplay between the fluid dynamics of the fuel/air inflows and the combustor acoustics under the operating conditions of flame B [55]. This mechanism introduces a different fuel-air mixing pattern at the nozzle exit, and this is also reflected in the downstream temperature and fuel mass fraction distributions shown in **Figure 2(b)** and **(c)**. Such mutually-interacting flow and flame dynamics cannot be captured by the conventional RANS modelling paradigm, which highlights the important role that LES can potentially play in jet engine combustor design and development.

To obtain the experimental time-averaged statistics shown in **Figure 2**, a 5-Hz laser system was used [50, 51] and no time-resolved measurement was available at the time. High-speed laser facilities have advanced rapidly in recent years allowing for measurements taken at a repetition rate up to several tens of kilo-Hertz. This hardware advancement has a significant impact on turbulent combustion experiments because the large-scale structures can now be readily resolved by the measurements in both physical space and time. From a modelling perspective, these measurements largely enrich the validation data and make it possible to assess the model capabilities of capturing transient behaviours. For the present combustor, for example, the dynamic motion of the coherent vortical structure, the so-called precessing vortex core (PVC), and the thermo-acoustic oscillation (TAO) were both identified using 10-kHz PIV measurements [52, 56]. **Figure 3** shows a typical comparison of measured and computed axial velocity spectra for two representative monitoring points (marked in **Figure 1a**) in flames A and B. These two points are located in the swirling jet and inner shear layer (between the jet and inner recirculation zone) respectively. The velocity spectra show strong dependence on the location and also behave quite differently in the two flames. The pronounced peaks correspond to the dominant frequencies for the PVC and TAO. These frequencies with their respective amplitudes are captured reasonably well in the simulation despite a considerable under-estimation of the PVC frequency for flame A. This suggests that the LES modelling framework and the FlaRe combustion sub-model described in Section 2 are adequate to capture the complex flow/flame/acoustic dynamics in this dual-swirl combustor, which are similar to those occurring in real gas turbine combustors.

As modern jet engines operate at fuel lean conditions to achieve low emissions, the flame is prone to local extinctions or even complete blow-off in the worst scenario. Such events are disastrous when occurring at high altitudes, where the engine is difficult to relight due to the low temperature, pressure and O_2 level environment. Therefore, the physical mechanism driving flame blow-off deserves a better understanding, which helps to develop not only control strategies but also predictive CFD tools for the engine design. To this end, the approaching blow-off flame C in **Table 1** was investigated experimentally by Stöhr et al. [54]. It was observed that the flame was highly unstable exhibiting sudden lift-off events with vanished flame root. Recently, this flame has been tackled using LES with FlaRe [57] and CMC [58] subgrid combustion closures, both showing a good agreement between the simulation and experiment for the flow and flame statistics. Although the flame root dynamics associated with the PVC motion was captured by both combustion models, the extreme lift-off event was only shown in the study using FlaRe [57]. This is probably due to the limitation of non-premixed CMC with a single conditioning variable-mixture fraction, while the flame root experiences strong partially premixing effects during lift-off [59]. To illustrate this lift-off event in a clear manner, **Figure 4** depicts the typical computed (using FlaRe [57]) and measured [54] flame roots for a stable instant at $t = 0$ ms and an extinguished instant at $t = 40$ ms. Despite the qualitative nature of this comparison, the simulated flame root behaviour agrees quite well with that measured using high-speed OH-PLIF and details can be found in [57]. Remarkably, the flame root extinction is successfully captured by an unstrained flamelet model, which suggests that the subgrid straining effect is not of leading order in the extinction process. This has a further implication that the cost-effective flamelet models can be used for

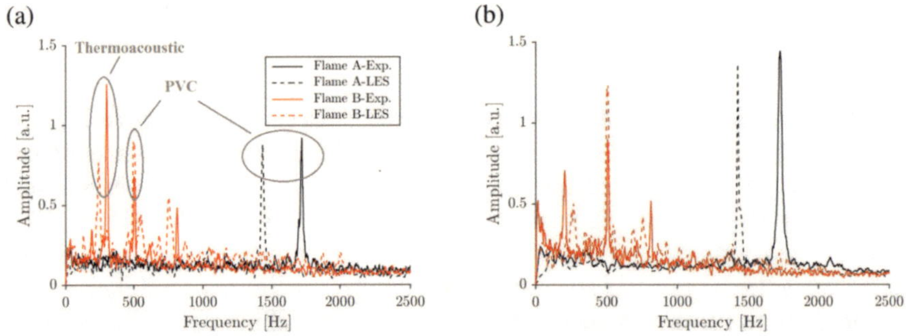

Figure 3.
*Comparison of axial velocity spectra for two monitor points (marked in **Figure 1a**) located in: (a) swirling jet and (b) inner shear layer near the nozzle exit.*

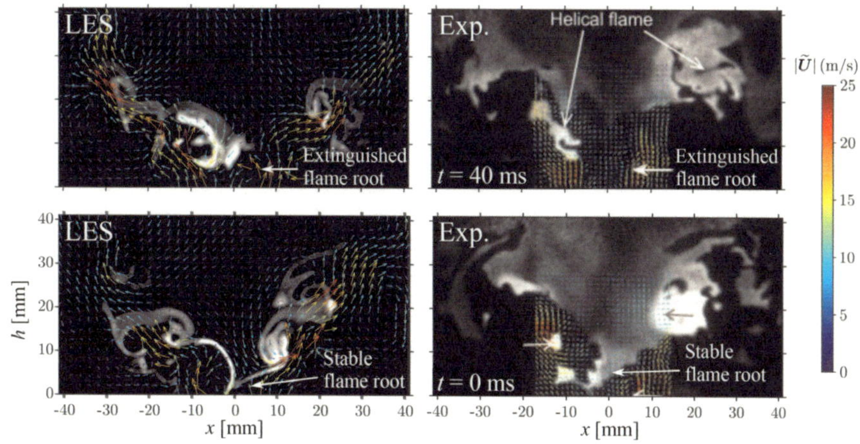

Figure 4.
LES [57] and experimental [54] snapshots of flame root overlaid by velocity arrows (coloured by magnitude) for typical stable (lower row) and extinguished (upper row) instants. The computed and measured flame is illustrated by filtered reaction rate contours and OH-PLIF images respectively.

prediction of blow-off, which itself is a slow (usually hundreds of ms) and hence computationally expensive process to simulate for practical combustor conditions.

3.2 Multi-burner annular combustor

In most modern jet engines, multiple burners are aligned circumferentially to achieve high thermal power within a compact volume. As a result, the unsteady heat release of these individual flames interact with the annular geometry of the combustion chamber, which gives rise to self-excited azimuthal instabilities [60]. Compared to longitudinal modes observed for a single flame, e.g., flame B of the DLR dual-swirl burner discussed earlier in this section, azimuthal modes are more dominant and destructive in practical applications because the circumference is usually shorter than the longitudinal length of the combustor resulting in higher resonant frequencies [61]. Thus, azimuthal instability is recognised as a primary issue for jet engine manufacturers.

Due to the complexity and high cost, only few laboratory model annular combustors have been studied experimentally so far, e.g., [62, 63], and the numerical works are scarce. To bridge this gap and gain physical insight into azimuthal instabilities, the annular burner of Worth et al. [62, 64] is simulated using the FlaRe model in this subsection. A photograph along with the schematic of this burner is shown in **Figure 5**. Fully premixed ethylene-air mixture was supplied at the bottom of the plenum and then passed through a honeycomb flow straightener before splitting into 12 bluff-body tubes by a hemispherical flow divider. Both swirling [62] and non-swirling [64] cases were investigated with and without the swirlers below the bluff-bodies. The experiments were operated at room temperature and atmospheric pressure. A bulk mean velocity evaluated at the bluff-body exit was kept constant at 18 m/s for all cases and pronounced azimuthal instability was observed for equivalence ratio ranging from $\phi = 0.8$ to 1.0. Three pressure transducers, denoted as P1, P2 and P3, were mounted on the tube wall 45 mm upstream of the bluff-body exit and they were separated by 120° to measure the azimuthal pressure waves travelling in the θ-direction.

The typical computed flame structures of the non-swirling and swirling cases for $\phi = 0.8$ are presented in **Figure 6** using volumetric rendering of the reaction rate for the 12 burners. The instantaneous axial velocity field is also shown for the mid-

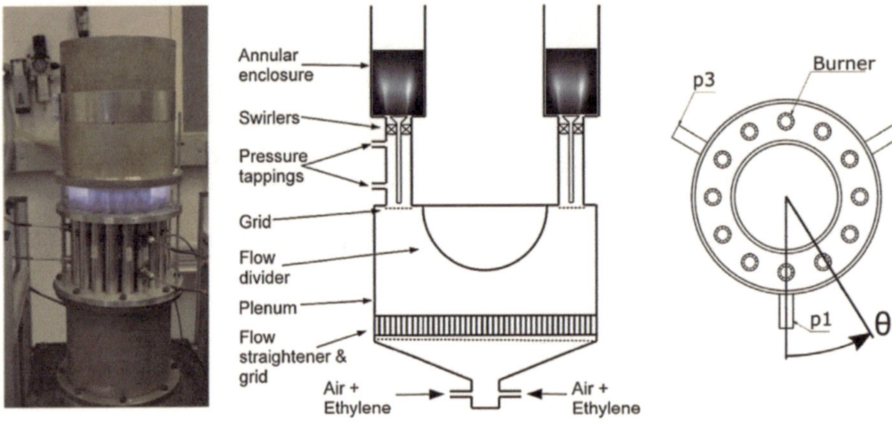

Figure 5.
Photograph and schematic of the annular combustor [62, 64].

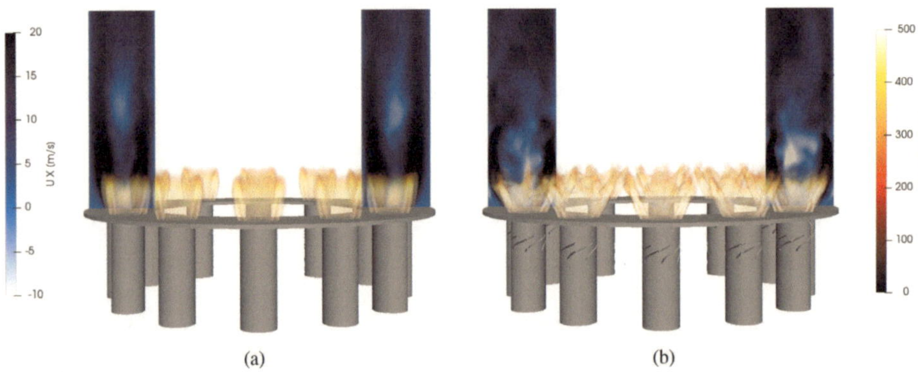

Figure 6.
Instantaneous reacting flow heat release volume rendering along with representative axial velocity contours for the (a) non-swirling and (b) swirling cases. Bulk mean velocity is $U_b = 18$ m/s with the equivalence ratio of $\phi = 0.8$.

plane of two opposed burners on the circumference. As expected, the swirl flames in **Figure 6(b)** are more compact having smaller flame lengths compared to the bluff-body flames in **Figure 6(a)**. Also, these flames are more opened up in the radial direction leading to shorter flame-to-flame distances. To qualitatively assess the LES results, **Figure 7** compares the measured and computed overhead line-of-sight integration of the mean heat release rate. In general, the distribution of the

Figure 7.
Overheard view of the line-of-sight (integrated in the axial direction) mean heat release rate for the non-swirling and swirling cases. Operating conditions are the same as in ***Figure 6***.

measured OH* signal, due to its more diffusive nature as light emissions, are more spread than the computed heat release rates for both cases with and without the swirlers. Otherwise the simulation results agree quite well with the measurements. The non-swirling flames are more or less symmetric across the annular chamber while the swirling ones clearly show a bulk swirl moving in the anti-clockwise (ACW) direction. This trend is qualitatively captured in the LES as seen in the figure.

Given that the good performance shown earlier for the single burner, it is not surprising that structure of the multiple flames is also well captured. The more important aspect of using LES for full-annular combustor is to examine its ability to predict the azimuthal instability, which is not present in single burners. This is of particular interest for the gas turbine industry because such a predictive tool which provides good accuracy at reasonable computational cost is highly needed but yet to be developed. **Figure 8** presents the typical pressure fluctuation time series taken at the three probes (marked in **Figure 5**) and their spectra for the non-swirling case with $\phi = 0.9$ and $U_b = 18$ m/s. A very clear azimuthal wave motion is seen as the p' signals of P1, P2 and P3 are exactly 120° out of phase. The computed frequency of this mode is about 1950 Hz, which agrees very well the measured value of about 1920 Hz. The small difference in the frequency could result from the adiabatic wall conditions assumed in the LES, leading to a higher speed of sound than that in the experiment with wall heat losses.

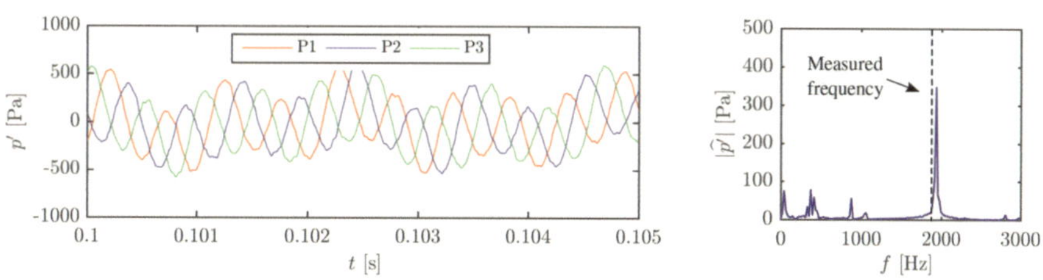

Figure 8.
Pressure fluctuation time series (left) and power spectra (right) for the non-swirling case with $\phi = 0.9$ and $U_b = 18$ m/s. the pressure probe locations are marked in ***Figure 5****.*

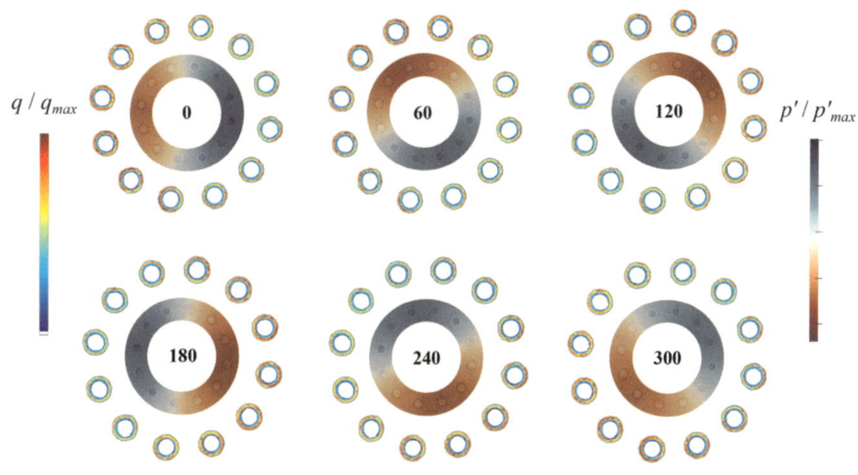

Figure 9.
Six phase angles of the azimuthal ACW spinning mode and phase-averaged heat release rates (integrated in the streamwise direction) in the transverse plane.

The pressure-heat release coupling is an essential mechanism for thermo-acoustic instabilities to occur. Thus, phase-averaged results are informative and often used to study modal behaviours of the instabilities. **Figure 9** shows the transverse-plane pressure oscillation and phase-averaged heat release rate at six different phase angles spanning over a thermo-acoustic cycle. The heat release contours are integrated values in the streamwise direction. It can be clearly seen that there is a substantial azimuthal variation in both the pressure and heat release fields. These fluctuations are strongly in phase and their peak magnitudes spin along the combustion chamber annulus at the speed of sound as also seen in the experiments [64]. However, unlike in the experiments, there is no mode switching behaviour (change of spinning direction or to standing mode) observed within the duration of the LES for this case. Such modal dynamics were detected at very low frequencies (less than 5 Hz) which require excessively long simulation runtime. This aspect is still beyond the capacities of the current petascale high-performance computing and will probably become possible in the near future as we approach the exascale era. Nonetheless, these results demonstrate that the FlaRe modelling framework used can accurately capture the major characteristics of azimuthal mode instability in annular combustors without any tuning of the model parameters. This successful LES exercise is among the firsts of its kind for the self-excited azimuthal instability in a full-annular combustor, and it is only possible when a robust, accurate and computationally inexpensive combustion model is appropriately coupled with the CFD solver. This modelling framework validated using laboratory cases in this section is readily applicable for practical combustors, which are discussed next.

4. Practical combustors

Practical combusting devices operate at high pressures, which can range from few bar for a compact power plant combustor to 30 or 40 bar for an aero engine at take-off conditions, with shaft power of the order of 100 kW to Megawatts per combustion sector. Higher powers are achieved using multi-sector and/or annular configurations. The need for high pressures lies in the efficiency of the Brayton cycle, which is the thermodynamic cycle that represents the functioning of a gas turbine. The operating principle is simple and is represented on the temperature-entropy plane in **Figure 10(a)**. The same cycle is sketched using the gas turbine components in **Figure 10(b)**. Combustion happens between points 2 and 3 of the

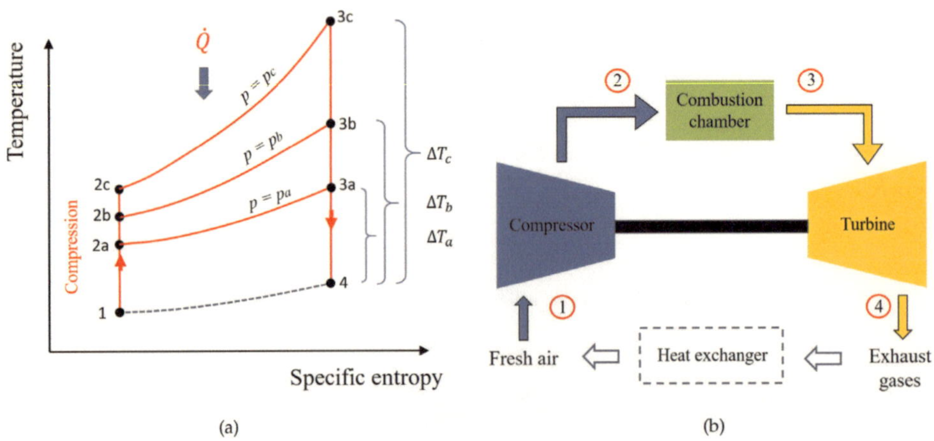

Figure 10.
Typical Brayton thermodynamic cycle for gas turbine: (a) temperature-entropy diagram and (b) sketch of the cycle components.

cycle, where chemical energy is converted into thermal energy. The gas pressure remains almost constant during this process and thus the corresponding curves on the $T-s$ plane are isobaric, and they diverge from each other as the initial temperature increases (points 2a to 2c in **Figure 10a**). This increase of temperature at point 2 of the cycle is achieved by corresponding increase in pressure by the compressor. The energy gain in terms of thermal energy can be quantified for a perfect gas by the variation of sensible enthalpy $dh_s = C_p dT$. Because of the divergence of the isobaric curves, an increase of temperature (or equivalently pressure) at the combustor entrance results in larger and larger gain of temperature and thus thermal energy at the combustor exit, point 3. For example, increasing the temperature of a quantity ΔT_i from point 2a to point 2b, results in a temperature gain $\Delta T_b - \Delta T_a > \Delta T_i$ at the combustor exit, i.e., the energy gain is larger than the amount that the turbine has to absorb to allow the compressor to yield the initial temperature increase. In other words, the higher the initial temperature (thus the pressure at point 2), the higher the energy gain, given by the thermal energy in output less the part needed by the turbine to run the compressor. There are however technological limitations for which the pressure at point 2 cannot be increased over a certain threshold, neither the temperature at point 3 can surpass a value given by structural limitations of the turbine blades, and these limitations are not going to be discussed here. It is worth mentioning that in the above discussion: (i) the cycle is ideal, i.e., irreversible losses were not taken into account; and (ii) the energy conversion between points 4 and 1 is conceptually represented by a heat exchanger and an equivalent isobaric curve. This can actually be present in a power plant where gases are recycled, but is only nominal in an aero engine, where the exhaust gases leave the system.

The design of high-pressure devices is complicated by the difficulty of having accurate measurements, in particular for temperature, on which the design process strongly relies. Non-intrusive laser techniques like Raman or Rayleigh scattering are very expensive at high pressures, and additional challenges exist because of safety reasons associated to creating an optical access in the pressurised combustion chamber area [65]. Moreover, sophisticated laser diagnostic techniques such as coherent anti-Stokes Raman spectroscopy (CARS), Raman or Rayleigh scattering may become less reliable for high pressure conditions. Several challenges exist also in numerical simulations. First, validation data from experiments is very limited for the reasons above, and this slows down the process of developing robust CFD tools to be used for the design process. Secondly, the flame thickness decreases by one or two order of magnitude as the pressure increases, about 1/10 or 1/100 of a millimetre. Given the complex geometry of modern combustion systems and their dimension which is of the order of tens of centimetres, it follows that to fully capture the small-scale combustion processes in a 3D CFD simulation the numerical grid becomes of order of hundred of millions cells. This is challenging for industrial purposes, where results are expected in order of days, despite the recent advances in high-performance computing technology, and even unaffordable when unsteady phenomena such as combustion instabilities are present, and relatively fast methods like RANS cannot be used or are unreliable. Unfortunately these instabilities are of paramount importance and their behaviour has to be understood before lean-operating, new generation engines can be developed. In this scenario it is clear that:

1. CFD modelling and in particular subgrid modelling for LES assumes a critical role to compensate for the experimental limitations and at the same time provide answers to the behaviour of unsteady phenomena such as combustion instability and local extinctions occurring in developmental combustion systems. The role of the turbulence-combustion interaction modelling is even

more critical to reduce to a minimum the mesh size and thus the simulation cost and runtime. In fact, reducing the mesh size to values of industrial practicality (order of 10 million or less) unavoidably implies that the local cell size is of order or larger than the flame thickness. It follows that the small scale processes have to be entirely modelled, which emphasises the role of the SGS modelling on the final results;

2. For a fixed mesh size, the CFD modelling has to be computationally fast. This drives the industrial choice towards specific types of modelling. In particular, flamelet-like models have attracted the interest of industries such as Rolls-Royce and Siemens for their advantages in terms of computational time (see Section 2). The limitations associated to flamelet assumption, however, lead to the need of further model development before this type of modelling can be effectively employed for design purposes.

The following subsections illustrate advantages and limitations of flamelet modelling for high pressure configurations in lean combustion systems, in light of recent CFD advancements. This is first shown for a power plant gas turbine operating at moderate pressure, where a good set of measurements and data from different combustion modelling is available for comparison. Then higher pressure configurations of aeronautical relevance are shown. These cases are chosen as they provide some limited but valuable experimental data for validation purposes.

4.1 Siemens combustor for energy generation

The following combustor sector is a modified version of the commercial SGT-100 family of Siemens, which consists of 6 combustors delivering a nominal shaft power of 5.7 MW. Each combustor burns natural gas after mixing with air in the swirler and prechamber of the geometry, shown in **Figure 11(a)**. The burner operates at 3 bar pressure, which is above atmospheric conditions, but is relatively low to allow a large database of in-flame measurements to be available for model validation, including temperature, velocity and major and minor species mass fractions radial profiles at four axial locations [67]. This configuration is swirled and features a PVC, which can be identified by looking at the velocity contours in the combustor primary zone in **Figure 11**. The stagnation point, marked in the figure, is in fact not on the centreline, suggesting that the PVC did not complete an entire

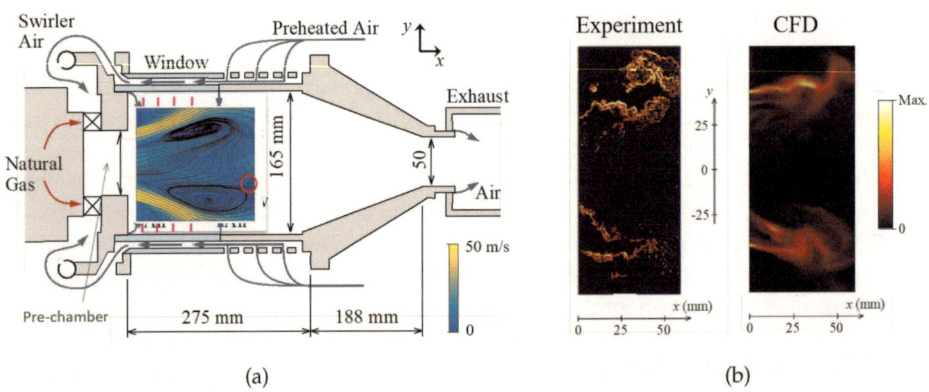

Figure 11.
Non-to-scale representation of the Siemens SGT-100 combustor with dimensions and velocity magnitude contours and streamlines from PIV measurements [22, 66] in the primary (combusting) zone) (a). The experimental and numerical flame are shown in (b) for a random instant of time.

revolution, and this has an effect on the statistics. The existence of a PVC is common in lean combustion burners and thus has to be taken carefully into account before comparisons are made with CFD results. Also, the PVC is usually coupled with the system acoustics, although this will not be discussed in the merit for this case. In addition to the PVC, the central recirculation zone, represented by the two large vortices in the figure, is strongly dependent on the jet angle at the exit of the prechamber, which thus affects the axial position of the stagnation point and, in turn, the statistics. These characteristics make the Siemens SGT-100 combustor a challenging case for model validation, which is useful to understand model advantages and limitations. These are discussed in light of recent modelling advances next.

The Siemens configuration has been investigated numerically using different LES combustion modelling, including TPDF/ESF [68], partially stirred reactor (PaSR) [69], TF [70, 71], FlaRe [22], eddy dissipation, fractal and approximate decomposition models [71]. Comparisons among different modelling techniques are also shown in [22, 71]. The combustion conditions for the Siemens configuration were noted in [67] to lie between thin and distributed reaction zones regimes of the turbulent combustion diagram [72]. According to this, the smallest turbulent eddies are able to penetrate the internal flame structure, thus invalidating the flamelet hypothesis. However, detailed interrogation [69] of the measured OH suggested that there were flamelets embedded in an environment of distributed combustion, i.e., flamelet structures and thus flamelet modelling are still possible at high Karlovitz number regimes, which was observed also for other configurations [36]. The Siemens configuration is thus a critical case as it opens the way to exploit the strong computational advantage of flamelets for highly turbulent, high pressure configurations typical of practical burners. As discussed earlier, there are two ways of proceeding to simulate a high pressure flame. One way is to decrease the cell dimensions (thus increase the mesh count) so that at least 5–10 numerical cells lie within the flame thickness and consequently a good part of the turbulence-flame interaction is captured at the resolved level in the LES. This decreases the impact of the SGS modelling on the statistics. Nevertheless, as explained earlier this is unpractical. The second approach is to have a coarser, affordable mesh size, with the SGS modelling playing a strong role. As combustion is a small scale process, this strongly reflects on the statistics, which is illustrated in **Figure 11(b)**. As the mesh is not fine enough to enter the flame structure, the numerical flame appears smoothed and filtered in respect to the experimental one, where the wrinkling effect of small vortices is observable. The big challenge is thus to have a modelling which, despite the inability to represent this at the resolved level, is able to capture the effects on a number of statistics (first and second moments, PDFs, etc.) and remain computationally cheap at the same time. The simulation cost for the Siemens combustor starts from about 550 CPU-hour per ms of simulation for a flamelet model and can increase significantly depending on modelling and grid size, although precise values were not reported for other combustion models used for the same configuration [68–71].

The performance of the FlaRe model discussed in Section 2 and its ability to predict the flow field characteristics can be assessed by comparing the CFD results to experimental data available for the Siemens configuration [67]. Typical comparisons of radial profiles of temperature and velocity are shown in **Figure 12** for two axial locations in the flame region (please refer to [22] for a full database of comparisons). The first location is about 19 mm downstream the pre-chamber exit, where the flow diverges due to the sudden expansion and the second is 70 mm further downstream, where the gases are close to burnt conditions. Velocity and its rms are predicted with good accuracy by the LES at the upstream position, but some mis-alignment of the peak values is observed for the mean velocity at the

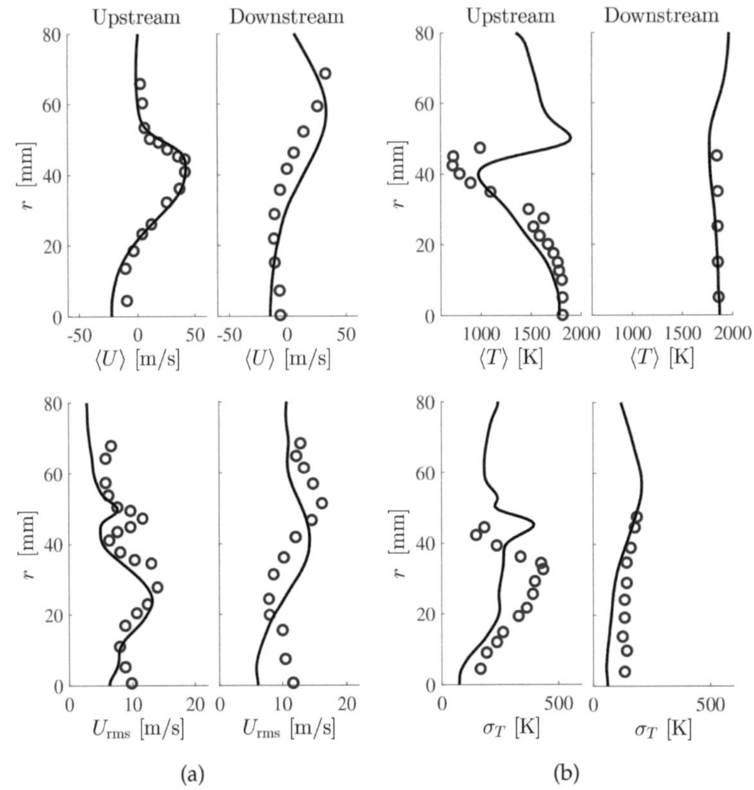

Figure 12.
Radial profiles of mean axial velocity, U (a), temperature, T (b), and their rms values at two axial positions in the primary zone of the SGT-100 combustor. Measurements (circles) are compared with LES results using FlaRe approach.

downstream position, which in turn affects the rms field. This is partly due to the fact that the LES is slightly over-predicting the jet angle at the combustor entrance (see **Figure 11**), which is also observed in predictions from other combustion models. Temperature profiles at the most upstream position show that the LES-FlaRe approach captures this quantity satisfactorily except for some over-prediction at $r \approx 40$ mm. This is the region where the flame anchors and is subjected to strong effect of strain [70]. It is possible that the grid resolution at this location needs further refinement to capture this effect at the resolved level. Similar over-predictions were observed using the TF approach with a similar grid resolution in [71]. While an improved accuracy was shown for TF model in [70] using 120 M cells increasing non-negligible computational cost, this may not be affordable for routine in-house calculations in industries. The work in [69] using PaSR model also shows that chemistry and in particular extinction strain rates may also play an effect at the same radial location (see also discussion in [22]). At the downstream locations where gases are burnt the temperature is predicted very well by the LES-FlaRe model, which is also a consequence of the fact that flamelet models guarantee that the correct adiabatic value is approached in burnt conditions, which may not be true for other modelling approaches. This is particularly relevant for real engines configurations where correct predictions of temperature and composition at the exhaust are needed for design purposes. The temperature rms also is satisfactorily predicted. It is worth mentioning that the heat released at the SGS scales has a strong effect on temperature and thus the portion of SGS temperature variance is large as compared to the resolved variance. Further modelling development is necessary to predict temperature fluctuations at SGS level. Prediction of temperature variance may become particularly relevant in situations where the measurement data is not density weighted (not the case for the Siemens combustor shown here, where measurement data is density weighted). As the reacting Navier-Stokes equations are density weighted, in such a case the LES data should be processed to obtain non-weighted averages using an approximation. This approximation will involve the estimation of the total variance (resolved plus SGS), as shown for example in [20]. The development of modelling to account for the SGS temperature variance in the statistics deserves thus a larger attention than that demonstrated in recent years.

Overall the flamelet model predictions are satisfactory and of similar accuracy than those obtained by other modelling approaches, at a significantly lower computational cost. Increase in accuracy can be achieved either increasing the mesh size (resolving more and more of the small-scale turbulence-flame interaction processes) or acting on the chemical mechanism (see discussion in [22] for more details), with different modelling giving similar performance at equal conditions of mesh and chemistry resolution. The recent advances in modelling development and in particular the progresses in the turbulence-reaction-dissipation balance have allowed flamelets to cover the gap that separated them from other modelling approaches. Also, the fully detailed mechanism used by flamelet models potentially allows to have information on more chemical species than in other models at no additional cost, as long as the correct flame-turbulence interaction is predicted. Note that this still does not imply that flamelet will be successful at higher pressures as the Siemens case clearly indicates that limitations exist in all combustion modelling when the LES filter size is larger than the flame thickness. The following subsection will shed some light on this.

4.2 Aero engine configurations

The lack and cost of experimental data, and the limitations of most combustion models to simulate complex high-pressure configurations in times affordable by industry, have slowed down the process of development of lean combustor technology. Flamelets models are computationally cheap enough to be used in industry but up to recently they have not been considered sufficiently accurate to be employed for high turbulence, high Karlovitz conditions for gas turbine combustion. The recent advances in flamelet modelling in the context of LES and the better understanding of the small-scale interaction between turbulence, reaction and diffusion as discussed in the previous sections, have shown potential to overcome the limitations of flamelets modelling and thus open the way to a faster design process.

In aero engines there is an additional modelling issue to consider which is due to the liquid fuel, usually kerosene or similar, which brings the modelling of the two-phase flow, fuel droplets break-up and their evaporation into consideration. These brings additional parameters and degrees of freedom in the CFD modelling and thus measurements of spray statistics such as Sauter mean diameter (SMD) and droplet velocity are needed to reduce the uncertainty in comparing CFD and experimental data. The spray behaviour, in general, both affects and is affected by the velocity and temperature field and thus it is not simple to separate spray and reaction effects. The validation of CFD models in aero engines thus leads to different considerations depending on whether the investigated region is close to the injectors or not.

4.2.1 Comparisons in the primary zone

The following test case is representative of a single sector aero engine combustor, where the spray statistics were observed to be only slightly affected by the

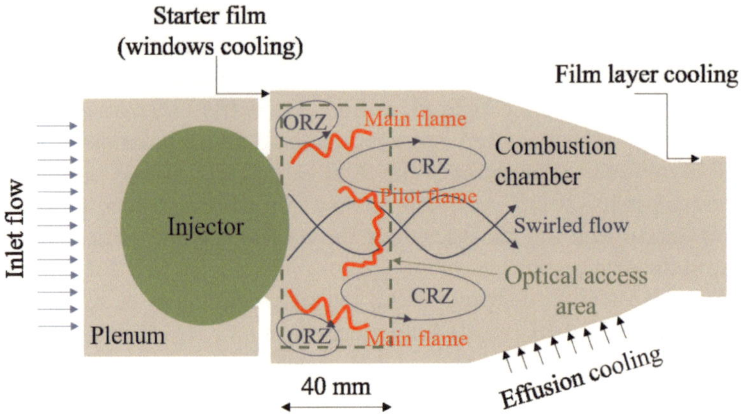

Figure 13.
Sketch of the pressurised BOSS rig of DLR operated with a Rolls Royce fuel injector.

surrounding field. A sketch of this combustor is shown in **Figure 13**. Pressurised air in the order of 10 bar and preheated at temperature T_{30} of about 700 K flows through a burner which consists of two stages: a central, pilot stage and a surrounding main stage. There are swirlers in each of these passages. These flow paths are designed to deliver different flow splits, and can have different channels of co-rotating or counter-rotating flows depending on the particular injector geometry and configuration. Liquid fuel is injected before the combustor entrance from the injector edges and is also split into pilot and main stages. The fuel split depends on the desired power settings (take-off, approach, idle, etc.). Correspondingly, the flame also consists of pilot and main branches, respectively stabilised in the internal part of the central recirculation zone (CRZ), and between the CRZ and the outer recirculation zone (ORZ) forming as consequence of the sudden expansion of the swirling flow at the chamber entrance. Film and effusion cooling are used to protect the walls from the high temperature gases in both primary and secondary zones. Previous studies [23] have shown that the spray statistics are not strongly influenced by the surrounding flow field in this configuration, so this case offers a good opportunity to evaluate the combustion model performance independently of the spray modelling. The computed SMD and droplet velocities were shown to compare well with measurements in [23]. When it comes to compare fields like temperature, the difficulties in having reliable measurements in the flame region lead, in the few cases where measurements are available at high pressure conditions, to significant uncertainties and this slows down the validation process of CFD models. For the studied configuration, direct measurements of OH concentration are available, with an uncertainty of 20–30% [73]. As other intermediate species, OH can be used to have a qualitative picture of the flame configuration and thus this quantity is still valuable for CFD validation purposes. Typical comparisons of OH mass fraction with LES-FlaRe predictions are shown for the primary zone in **Figure 14a**. Qualitatively, the LES-flamelet approach shows to be able to predict the correct flame configuration, involving a penetrating pilot (central) jet. This is challenging as an incorrect balance of reaction and turbulence can result in a completely different configuration with a diverging pilot jet and a flame anchored upstream in a V-shape [23]. Quantitatively, the OH concentration from the LES can over-predict that from experiment of a factor of two or larger as observed in the figure. However, this has to be carefully interpreted due to the uncertainty in the measurement and considering that the real objective is to predict temperature. This has a favourable non-linear dependence on OH (OH increases exponentially with

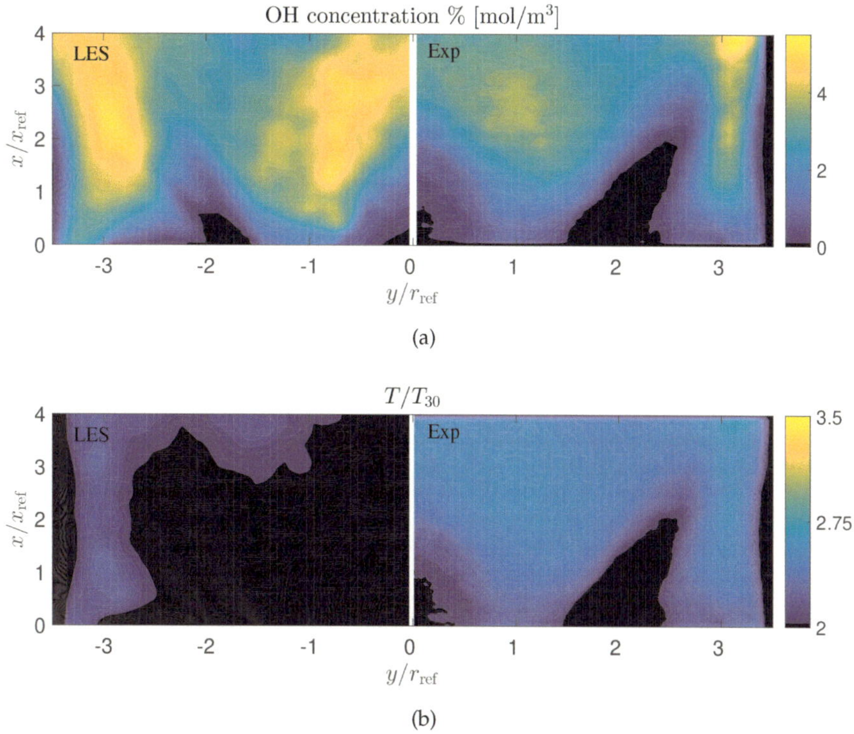

Figure 14.
Comparison of mean OH concentration (a) and temperature (b) from LES and measurements (for temperature) in the mid-plane of the primary zone of the Rolls-Royce Boss-rig. The temperature is normalised by the inlet temperature, T_{30}, for confidential reasons.

temperature) and thus the differences observed in **Figure 14a** are expected to become much smaller in terms of temperature. Unfortunately, direct measurements of temperature at high pressure condition are challenging as explained earlier and this quantity is if often estimated indirectly by making additional assumptions, which in turns lead to additional uncertainty. For example, for the combustor test case investigated here temperature can be estimated from OH concentration via equilibrium assumption [73]. A direct comparison of the experimental temperature in this case with that obtained in the LES from Eq. (7) can lead to incorrect conclusions if the underlying assumptions used in the experimental data are not carefully taken into account. An example of this is shown in **Figure 14b**, where the LES temperature appears to be significantly under-estimated in respect to that from experiments in the pilot flame region. This would be inconsistent to the behaviour observed for **Figure 14a** and suggests that comparisons of temperature in the burner primary region have to be assessed with due care at high pressure conditions.

4.2.2 Comparisons at the combustor exit

Comparisons between LES and measurements are more meaningful at the combustor exit where the gases are almost entirely combusted and thus assumptions such as that of chemical equilibrium are expected to better hold. Also, measurements at the combustor exit are as important as those in the primary region as the flow field here is the result of what happens upstream. Thus, experimental data at the combustor exit can be used for model validation with an increased degree of

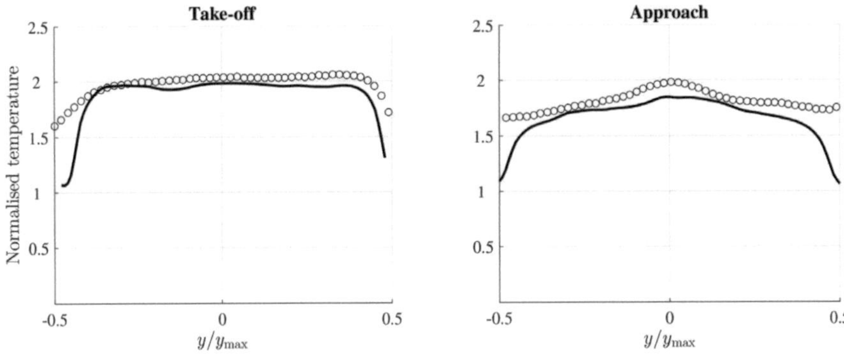

Figure 15.
Comparison of temperature profiles from measurements (symbols, courtesy of DLR Cologne, Germany) and LES (lines) at the exit plane of the DLR OCORE-2 rig of a practical single-sector aero engine combustor for two operating conditions.

quantitativeness in comparison to the primary zone. Measurements at the burner exit are not available for the configuration investigated in the previous section; however, temperature measurements are available for a similar rig, featuring a similar injector and the same flow configuration of **Figure 13**. Comparisons between experimental data and FlaRe-LES results are shown in **Figure 15** for two operating conditions at the same pressure and inlet temperature, but different flow split. These configurations are representative of approach and cutback conditions of an airplane. The differences observed near the walls of the combustor ($y/y_{\max} = \pm 0.5$) are due to the effusion cooling that lowers the temperature below the minimum detectable from the experiment (about 1200 K). Except for this region, the FlaRe model prediction matches very well that from experiment, which shows that this type of modelling is capable to represent the correct statistical behaviour even at high pressure when the intricate balance between turbulence, dissipation and heat release is correctly taken into account. Recent advances in the modelling in context of flamelets are thus promising for future design cycles of aero engines, although additional validations are still needed.

5. Summary and future outlook

In this chapter, an overview for the current status of the use of combustion CFD in modern gas turbine engine combustor design is presented. There is a general tendency in the industry to move from the conventional RANS to the more powerful LES modelling paradigm, and thus the discussion is focused on the application of LES. The various challenges for LES modelling of gas turbine combustion are discussed and a number of representative subgrid combustion models are briefly described. Flamelet approaches are more attractive for industry because of their significantly higher computational efficiency and relatively simple implementation in different CFD codes. The particular focus was given to a recently developed model called FlaRe, which is a revised flamelet approach keeping the physical consistencies among various SGS models and physical processes. To assess the performance of FlaRe, the LES results are compared with experimental measurements for several typical laboratory and practical combustors. A broad range of phenomena of high practical interest are involved in these test cases including flame-vortex interaction, self-excited thermoacoustic oscillations, flame root dynamics close to lean blow-off, high pressure conditions, liquid fuel combustion, etc.

The combustion regimes involved span over the full range for practical jet engine conditions involving premixed, non-premixed and a mixture of both. An overall good agreement between simulation and experiment is observed across all cases presented. This suggests that despite the limitations of the fundamental flamelet concept, which many believe is far from being valid for real industrial conditions, there is a great potential for flamelet models to be used in the industry on a frequent basis because of its computational efficiency, robustness and improved accuracy if the consistencies are maintained. This modelling framework is yet to be extended to cover other important aspects such as non-adiabatic effects, pollutant emission, autoignition, etc., of a real engine combustor.

Acknowledgements

The authors IL and NS thank the DLR Institute of Propulsion Technology in Cologne for kindly providing measurement data for some of the figures shown in this chapter, and in particular Dr C. Willert, Dr T. Behrendt and Dr J. Heinze for their useful advice on Section 4.2. The support from Mitsubishi Heavy Industries, Takasago, Japan is acknowledged by ZXC and NS. The presented simulations used the ARCHER UK National Supercomputing Service and the CSD3 Cluster of Cambridge University.

Author details

Zhi X. Chen[1]*, Ivan Langella[2] and Nedunchezhian Swaminathan[1]

1 Department of Engineering, University of Cambridge, Cambridge, UK

2 Department of Aeronautical and Automotive Engineering, Loughborough University, Loughborough, UK

*Address all correspondence to: zc252@cam.ac.uk

References

[1] Lefebvre AH, Ballal DR. Gas Turbine Combustion: Alternative Fuels and Emissions. Boca Raton, USA: CRC Press; 2010

[2] Poinsot T, Veynante D. Theoretical and Numerical Combustion. 2nd ed. Philadelphia, USA: Edwards; 2005

[3] Pope SB. Turbulent Flows. Cambridge, UK: Cambridge University Press; 2000

[4] Gicquel LYM, Staffelbach G, Poinsot T. Large Eddy simulations of gaseous flames in gas turbine combustion chambers. Progress in Energy and Combustion Science. 2012;**38**:782-817

[5] Pitsch H. Large-Eddy simulation of turbulent combustion. Annual Review of Fluid Mechanics. 2006;**38**: 453-482

[6] Smagorinsky J. General circulation experiments with the primitive equations. Monthly Weather Review. 1963;**91**(3):99-164

[7] Langella I, Doan NAK, Swaminathan N, Pope SB. Study of subgrid-scale velocity models for reacting and nonreacting flows. Physical Review Fluids. 2018;**3**:054602

[8] Swaminathan N, Bray KNC, editors. Turbulent Premixed Flames. Cambridge, UK: Cambridge University Press; 2011

[9] Colin O, Ducros F, Veynante D, Poinsot T. A thickened flame model for large eddy simulations of turbulent premixed combustion. Physics of Fluids. 2000;**12**:1843

[10] De A, Acharya S. Large Eddy simulation of a premixed Bunsen flame using a modified thickened-flame model at two Reynolds number. Combustion Science and Technology. 2009;**181**: 1231-1272

[11] Janicka J, Sadiki A. Large eddy simulation of turbulent combustion systems. Proceedings of the Combustion Institute. 2005;**30**:537-547

[12] Moureau V, Fiorina B, Pitsch H. A level set formulation for premixed combustion LES considering the turbulent flame structure. Combustion and Flame. 2008;**156**:801-812

[13] Pitsch H. A consistent level set formulation for large-eddy simulation of premixed turbulent combustion. Combustion and Flame. 2005;**143**: 587-598

[14] Dunn MJ, Masri AR, Bilger RW, Barlow RS, Wang GH. The compositional structure of highly turbulent piloted premixed flames issuing into a hot Coflow. Proceedings of the Combustion Institute. 2009;**32**: 1779-1786

[15] Poludnenko AY, Oran ES. The interaction of high-speed turbulence with flames: Global properties and internal flame structure. Combustion and Flame. 2010;**157**:995-1011

[16] Temme JE, Wabel TM, Skiba AW, Driscoll JF. Measurements of premixed turbulent combustion regimes of high Reynolds number flames. In: 53rd AIAA Aerospace Sciences Meeting. AIAA. Kissimmee, Florida: American Institute of Aeronautics and Astronautics. 2015

[17] Poinsot T, Veynante D, Candel S. Quenching processes and premixed turbulent combustion diagrams. Journal of Fluid Mechanics. 1991;**228**: 561-606

[18] Roberts WL, Driscoll JF, Drake MC, Goss LP. Images of the quenching of a flame by a vortex—To quantify regimes of turbulent combustion. Combustion and Flame. 1993;**94**:58-69

[19] Duwig C. Study of a filtered flame formulation for large eddy simulation of premixed turbulent flames. Flow, Turbulence and Combustion. 2007;**79**: 433-454

[20] Langella I, Swaminathan N. Unstrained and strained flamelets for LES of premixed combustion. Combustion Theory and Modelling. 2016;**20**:410-440

[21] Doan NAK, Swaminathan N, Chakraborty N. Multiscale analysis of turbulence-flame interaction in premixed flames. Proceedings of the Combustion Institute. 2016;**36**:1929-1935

[22] Langella I, Chen ZX, Swaminathan N, Sadasivuni SK. Large-Eddy simulation of reacting flows in industrial gas turbine combustor. Journal of Propulsion and Power. 2018;**34**: 1269-1284

[23] Langella I, Heinze J, Behrendt T, Swaminathan N, Zedda M, Voigt L. Turbulent flame shape switching at conditions relevant for gas turbines. Journal of Engineering for Gas Turbines and Power. 2019;GTP-19-1385

[24] Chen ZX, Langella I, Swaminathan N, Stöhr M, Meier W, Kolla H. Large Eddy simulation of a dual swirl gas turbine combustor: Flame/flow structures and stabilisation under thermoacoustically stable and unstable conditions. Combustion and Flame. 2019;**203**:279-300

[25] Bilger RW. Structure of diffusion flames. Combustion Science and Technology. 1976;**13**:155-170

[26] Demoulin FX, Borghi R. Modeling of turbulent spray combustion with application to diesel like experiment. Combustion and Flame. 2002;**129**: 281-293

[27] Giusti A, Mastorakos E, Hassa C, Heinze J, Magens E, Zedda M. Investigation of flame structure and soot formation in a single sector model combustor using experiments and numerical simulations based on the large eddy simulation/conditional moment closure approach. Journal of Engineering for Gas Turbines and Power. 2018;**140**:061506-061509

[28] Fiorina B, Gicquel O, Vervisch L, Carpentier S, Darabiha N. Approximating the chemical structure of partially premixed and diffusion counterflow flames using FPI flamelet tabulation. Combustion and Flame. 2005;**140**:147-160

[29] Bray KNC, Domingo P, Vervisch L. Role of the progress variable in models for partially premixed turbulent combustion. Combustion and Flame. 2005;**141**(4):431-437

[30] Bradley D, Gaskell PH, Lau AKC. A mixedness-reactedness flamelet model for turbulent diffusion flames. Proceedings of the Combustion Institute. 1990;**23**(1):685-692

[31] Salehi MM, Bushe WK, Shahbazian N, Groth CPT. Modified laminar flamelet presumed probability density function for LES of premixed turbulent combustion. Proceedings of the Combustion Institute. 2013;**34**:1203-1211

[32] Domingo P, Vervisch L, Payet S, Hauguel R. DNS of a premixed turbulent V flame and LES of a ducted flame using FSD-PDF subgrid scale closure with FPI-tabulated chemistry. Combustion and Flame. 2005;**143**: 566-586

[33] Landenfeld T, Sadiki A, Janicka J. A turbulence-chemistry interaction model based on a multivariate presumed Beta-PDF method for turbulent flames. Flow, Turbulence and Combustion. 2002;**68**: 111-135

[34] Lecocq G, Richard S, Colin O, Vervisch L. Hybrid presumed pdf and

flame surface density approaches for large-eddy simulation of premixed turbulent combustion: Part 1: Formalism and simulation of a quasi-steady burner. Combustion and Flame. 2011;**158**: 1201-1214

[35] Vreman AW, van Oijen JA, de Goey LPH. Subgrid scale modelling in large-eddy simulation of turbulent combustion using premixed flamelet chemistry. Flow, Turbulence and Combustion. 2009;**82**:511-535

[36] Langella I, Swaminathan N, Pitz RW. Application of unstrained flamelet SGS closure for multiregime premixed combustion. Combustion and Flame. 2016;**173**:161-178

[37] Chen ZX, Doan NAK, Ruan S, Langella I, Swaminathan N. *A priori* investigation of subgrid correlation of mixture fraction and progress variable in partially premixed flames. Combustion Theory and Modelling. 2018;**22**:862-882

[38] Fritz J, Kröner M, Sattelmayer T. Flashback in a swirl burner with cylindrical premixing zone. In: International Gas Turbine and Aeroengine Congress and Exhibition. ASME. New Orleans, LA: American Society of Science and Engineering. 2001

[39] Schmehl R, Maier G, Wittig S. CFD analysis of fuel atomization, secondary droplet breakup and spray dispersion in the premix duct of a LPP combustor. Proceedings of the ICLASS. 2000

[40] Pera C, Réveillon J, Vervisch L, Domingo P. Modeling subgrid scale mixture fraction variance in LES of evaporating spray. Combustion and Flame. 2006;**146**:635-648

[41] Dunstan TD, Minamoto Y, Chakraborty N, Swaminathan N. Scalar dissipation rate modelling for large Eddy simulation of turbulent premixed flames. Proceedings of the Combustion Institute. 2013;**34**:1193-1201

[42] Ruan S, Swaminathan N, Darbyshire OR. Modelling of turbulent lifted jet flames using flamelets: *A priori* assessment and *a posteriori* validation. Combustion Theory and Modelling. 2014;**18**(2):295-329

[43] TNF Workshop;. Available from: https://www.sandia.gov/TNF/abstract.html

[44] Pope SB. Small scales, many species and the manifold challenges of turbulent combustion. Proceedings of the Combustion Institute. 2013;**34**:1-31

[45] Ahmed SF, Mastorakos E. Spark ignition of lifted turbulent jet flames. Combustion and Flame. 2006;**146**(1–2): 215-231

[46] Chen Z, Ruan S, Swaminathan N. Large Eddy simulation of flame edge evolution in a spark-ignited methane-air jet. Proceedings of the Combustion Institute. 2017;**36**:1645-1652

[47] Zhang H, Giusti A, Mastorakos E. LES/CMC modelling of ignition and flame propagation in a non-premixed methane jet. Proceedings of the Combustion Institute. 2019;**37**:2125-2132

[48] Lacaze G, Cuenot B, Poinsot T, Oschwald M. Large eddy simulation of laser ignition and compressible reacting flow in a rocket-like configuration. Combustion and Flame. 2009;**156**: 1166-1180

[49] Jones WP, Prasad VN. LES-pdf simulation of a spark iginited turbulent methane jet. Proceedings of the Combustion Institute. 2011;**33**:1355-1363

[50] Weigand P, Meier W, Duan XR, Stricker W, Aigner M. Investigations of swirl flames in a gas turbine model combustor. I. Flow field, structures, temperature, and species distributions.

Combustion and Flame. 2006;**144**: 205-224

[51] Meier W, Duan XR, Weigand P. Investigations of swirl flames in a gas turbine model combustor. II. Turbulence–chemistry interactions. Combustion and Flame. 2006;**144**: 225-236

[52] Stöhr M, Boxx I, Carter CD, Meier W. Experimental study of vortex-flame interaction in a gas turbine model combustor. Combustion and Flame. 2012;**159**:2636-2649

[53] Steinberg AM, Boxx I, Stöhr M, Carter CD, Meier W. Flow–flame interactions causing acoustically coupled heat release fluctuations in a thermo-acoustically unstable gas turbine model combustor. Combustion and Flame. 2010;**157**:2250-2266

[54] Stöhr M, Boxx I, Carter C, Meier W. Dynamics of lean blowout of a swirl-stabilized flame in a gas turbine model combustor. Proceedings of the Combustion Institute. 2011;**33**: 2953-2960

[55] Chen ZX, Swaminathan N, Stöhr M, Meier W. Interaction between self-excited oscillations and fuel-air mixing in a dual swirl combustor. Proceedings of the Combustion Institute. 2019;**37**: 2325-2333

[56] Boxx I, Stöhr M, Carter C, Meier W. Temporally resolved planar measurements of transient phenomena in a partially pre-mixed swirl flame in a gas turbine model combustor. Combustion and Flame. 2010;**157**:1510-1525

[57] Massey JC, Chen ZX, Swaminathan N. Lean flame root dynamics in a gas turbine model combustor. Combustion Science and Technology. 2019. DOI: 10.1080/00102202.2019.1584616

[58] Zhang H, Mastorakos E. LES/CMC modelling of a gas turbine model combustor with quick fuel mixing. Flow, Turbulence and Combustion. 2018. DOI: 10.1007/s10494-018-9988-1

[59] Lawn CJ. Lifted flames on fuel jets in co-flowing air. Progress in Energy and Combustion Science. 2009;**35**(1): 1-30

[60] Lieuwen TC, Yang V. Combustion Instabilities in Gas Turbine Engines: Operational Experience, Fundamental Mechanisms, and Modeling. Reston, VA: AIAA Inc; 2006

[61] Dowling AP, Stow SR. Acoustic analysis of gas turbine combustors. Journal of Propulsion and Power. 2003; **19**:751-764

[62] Worth NA, Dawson JR. Modal dynamics of self-excited azimuthal instabilities in an annular combustion chamber. Combustion and Flame. 2013; **160**:2476-2489

[63] Bourgouin JF, Durox D, Moeck JP, Schuller T, Candel S. A new pattern of instability observed in an annular combustor: The slanted mode. Proceedings of the Combustion Institute. 2015;**35**:3237-3244

[64] Mazur M, Nygard HT, Dawson JR, Worth NA. Characteristics of self-excited spinning azimuthal modes in an annular combustor with turbulent premixed bluff-body flames. Proceedings of the Combustion Institute. 2019;**37**:5129-5136

[65] Kohse-Höinghaus K, Barlow RS, Aldén M, Wolfrum J. Combustion at the focus: Laser diagnostics and control. Proceedings of the Combustion Institute. 2005;**30**:89-123

[66] Stopper U, Aigner M, Ax H, Meier W, Sadanandan R, Stör M, et al. PIV, 2D-LIF and 1D-Raman measurements of flow field, composition and temperature in premixed gas turbine flames.

Experimental Thermal and Fluid Science. 2010;**34**:396-403

[67] Stopper U, Meier W, Sadanandan S, Stör M, Aigner M, Bulat G. Experimental study of industrial gas turbine flames including quantification of pressure influence on flow field, fuel/air premixing and flame shape. Combustion and Flame. 2013;**160**: 2103-2118

[68] Bulat G, Jones WP, Marquis AJ. NO and CO formation in an industrial gas-turbine combustion chamber using LES with the Eulerian sub-grid PDF method. Combustion and Flame. 2014;**161**: 1804-1825

[69] Bulat G, Fedina E, Fureby C, Meier W, Stopper U. Reacting flow in an industrial gas turbine combustor: LES and experimental analysis. Proceedings of the Combustion Institute. 2015;**35**: 3175-3183

[70] Jaravel T, Riber E, Cuenot B, Buat G. Large Eddy simulation of an industrial gas turbine combustor using reduced chemistry with accurate pollutant prediction. Proceedings of the Combustion Institute. 2016;**36**: 3817-3825

[71] Fedina E, Fureby C, Bulat G, Meier W. Assessment of finite rate chemistry large Eddy simulation combustion models. Flow, Turbulence and Combustion. 2017;**99**:385-409

[72] Peters N. Turbulent Combustion. Cambridge, UK: Cambridge University Press; 2000

[73] Heinze J, Meier U, Behrendt T, Willert C, Geigle KP, Lammel O, et al. PLIF thermometry based on measurements of absolute concentrations of the OH radical. Zeitschrift für Physikalische Chemie. 2011;**225**:1315-1341

Chapter 4

Breakup Morphology and Mechanisms of Liquid Atomization

Hui Zhao and Haifeng Liu

Abstract

Fuel atomization, the transformation of bulk liquid into sprays, is of importance in jet engines. In this chapter, we will introduce the latest research advances on breakup morphology and mechanism of liquid atomization. On primary atomization, based on the morphological difference, the twin-fluid atomization could be classified into different regimes. The influence of Kelvin-Helmholtz and Rayleigh-Taylor instability on breakup morphology and fragment size is great and nonmonotonic. On secondary atomization, Rayleigh-Taylor instability is considered as the main driving mechanism in different bag-breakup modes; for higher Weber number, it will be in concurrence with the shear instability. Based on ligament-mediated spray formation model, ligament breakup is found to be well represented by the gamma distributions. Atomization of complex fluids has special characteristics and mechanisms. There are also a lot of research advances recently in this field.

Keywords: atomization, sprays, breakup, drop, instability

1. Introduction

Transformation of bulk liquid fuel into sprays is of importance in many engines. Most fuels employed in engines are liquid that must be atomized before being injected into combustion zone. Atomization could produce a very high ratio of surface to mass in the liquid phase, thereby promoting rapid reaction and combustion. In addition, liquid atomization is also common in a wide array of applications, such as agriculture, coatings, gasification, water scrubber, pharmaceuticals, metal powder production, 3D printing, spray drying, fire suppression, and cooling.

Atomization quality can be described in terms of mean drop size and distribution. Important factors in atomization include the flow conditions, liquid properties, gas properties, atomizer (or nozzle, injector) dimensions, and environment conditions. So, there are many atomizer types in the industry and laboratory, such as pressure atomizers, air-blast atomizers, air-assist atomizers, rotary atomize, effervescent atomizers, electrostatic atomizers, ultrasonic atomizers, etc. Then, the key is to select the suitable atomizers for the given application, which can have a well performance in operating conditions.

Atomization process usually consists of the initial removal of liquid mass from the surface to form large liquid drops and the subsequent breakup of these drops into tiny droplets; the phenomena are, respectively, known as primary and secondary atomization. Atomization has been quantitatively studied for more than a century. However, liquid atomization is a complicate, multiparameter two-phase flow process, which is not well understood. Many empirical theories and equations have been developed and used in atomization. So, there is still a lot of unknown work to be done [1–15].

2. Dimensionless number

The physical processes and fluid properties are important in atomization morphology and performance, and the mathematical and numerical analysis of atomization is very challenging, so a number of dimensionless numbers are used. First of all, the most important one is Weber number, which represents the ratio of disruptive hydrodynamic forces to the stabilizing surface tension force,

$$We = \frac{\rho_g \left(u_g - u_l\right)^2 D}{\sigma}, \tag{1}$$

where ρ_g is the gas density, u_g is the gas velocity, u_l is the liquid velocity, D is the characteristic size (in general, nozzle diameter in primary atomization and drop diameter in secondary atomization), and σ is the surface tension. Liquid viscosity will hinder deformation and dissipates energy supplied by aerodynamic forces. The viscosity effect is highly correlated with the Ohnesorge number,

$$Oh = \frac{\mu_l}{\sqrt{\rho_l D \sigma}}, \tag{2}$$

where μ_l is the liquid viscosity, and ρ_l is the liquid density. Other important dimensionless groups are gaseous Reynolds number

$$\mathrm{Re}_g = \frac{\rho_g u_g D}{\mu_g}, \tag{3}$$

liquid Reynolds number

$$\mathrm{Re}_l = \frac{\rho_l u_l D}{\mu_l}, \tag{4}$$

Mach number

$$Ma = u_g / c, \tag{5}$$

Strouhal number

$$St = \frac{fD}{u_l}, \tag{6}$$

the characteristic time [16]

$$T = \frac{t u_g}{D} \sqrt{\frac{\rho_g}{\rho_l}}, \tag{7}$$

gas-liquid momentum flux ratio

$$M = \frac{\rho_g u_g^2}{\rho_l u_l^2}, \tag{8}$$

momentum ratio

$$MR = \frac{\rho_g u_g^2 A_g}{\rho_l u_l^2 A_l} \tag{9}$$

and mass ratio

$$GLR = \frac{\rho_g u_g A_g}{\rho_l u_l A_l}. \tag{10}$$

Here μ_g is the gas viscosity, c is the speed of sound in the medium, f is the frequency, t is the real time, and A_g and A_l are the area of gas exit and liquid exit, respectively.

Complex fluid is usually a kind of complicated non-Newtonian fluid, which has more dimensionless groups on atomization. For example, in the Bingham model, the flow is characterized by the following constitutive equations,

$$\tau = \tau_0 + \eta\gamma \text{ and } \mu_l = \tau_0/\gamma + \eta, \tag{11}$$

where τ is the shear stress, τ_0 is the yield stress, η is the plastic viscosity, and γ is the shear rate. Hedstrom number is the useful nondimensional number, which depends only on material properties and geometrical parameters,

$$He = \frac{\tau_0 D^2 \rho}{\eta^2}. \tag{12}$$

The ratio of the yield stress to the stabilizing surface tension force is [17]

$$X = \tau_0 D/\sigma, \tag{13}$$

and the ratio of the aerodynamic force to the yield stress is [17]

$$Y = \rho_g u_g^2/\tau_0. \tag{14}$$

For viscoelastic fluids, Weissenberg number compares the elastic forces to the viscous forces

$$Wi = N_1/\tau \text{ or } Wi = t_R\gamma, \tag{15}$$

where N_1 is the first normal-stress difference, and t_R is the relaxation time. Another dimensionless number on the ratio of first normal stress difference to surface tension force could be defined as follows [18]

$$Z = \frac{N_1}{\sigma D}. \tag{16}$$

3. Primary atomization

Nozzle is generally used to produce spray. Fuel injection process plays a major role in many aspects of combustion performance. The influence of nozzle structure is remarkable on the atomization performance [19–22]. With the progress of technology

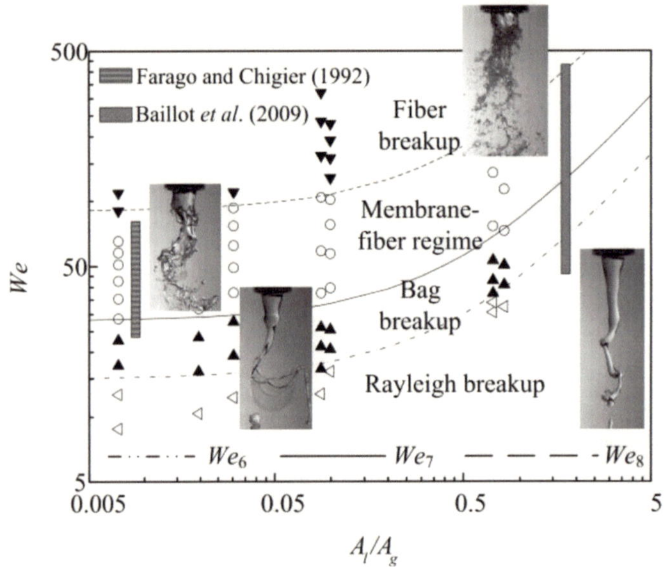

Figure 1.
Influence of A_g/A_l on atomization mode.

[23], it is found that the properties of breakup morphology and fragment distribution in different regimes are different [24–28]. Coaxial air-blast atomizers have many applications [29–31]. In order to obtain the desired results of atomization in the industrial scale, the suitable range of nozzle size and operating condition could be determined with the help of the regime map. There are two basic types on coaxial gas-liquid jets: (I) a cylindrical liquid jet surrounded by an annular gaseous stream and (II) an annular liquid sheet with an inner round gaseous stream [32–35].

In cylindrical liquid jet and annular gas jet, the common atomization modes are Rayleigh-type breakup (axisymmetric and non-axisymmetric), the membrane-type breakup (bag-type and membrane-fiber), fiber-type breakup, superpulsating breakup, atomization, and so on [36–42]. For the traditional classification, the

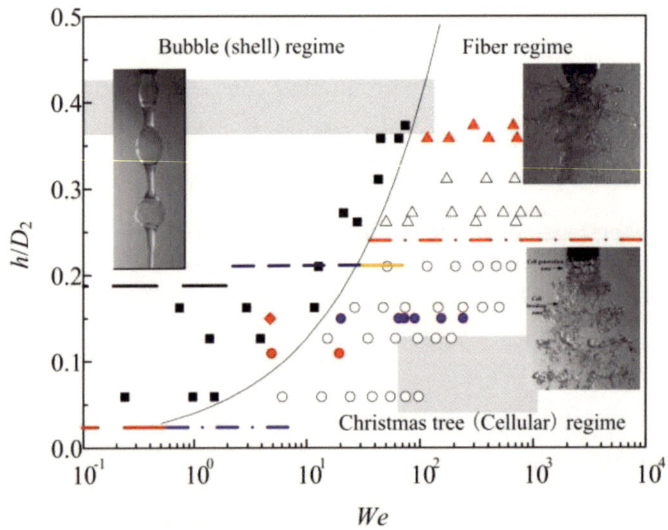

Figure 2.
Influence of h/D on atomization mode.

We-Re_l map or M-Re_g map is common. However, these criteria cannot reflect the impact of gas-liquid nozzle exit size. So, there is the modified map of classification by the gas-liquid nozzle exit area ratio A_g/A_l. **Figure 1** shows that atomization performance will improve with the increase of A_g/A_l, but the oversized value of A_g/A_l will produce very little effect and waste energy [41]. These results are conducive to the design of nozzles.

In cylindrical gas jet and annular liquid jet, the common atomization modes are bubble (shell) breakup, Christmas tree (cellular) breakup, fiber breakup, and so on [43–55]. Here, the characteristic size is the liquid film thickness h, which has an important impact on liquid film breakup. Then, there is the h/D-We map of breakup regimes based on the Rayleigh-Taylor instability is proposed, which is in well agreement with the experimental results as shown in **Figure 2**. Note that outlet wall thickness of nozzle can affect the flow field at nozzle outlet [56–61].

4. Secondary atomization

Drop is subjected to aerodynamic forces when there is the relative velocity between drop and gas. This force results in drop deformation and, if sufficiently large, will lead to breakup and fragmentation. Differing gas flow conditions can lead to differing drop breakup modes. Based on the morphology, as We increases, the vibrational, bag, bag-stamen, multimode, sheet-thinning, shear, and catastrophic breakup mode appear in turn [62–66]. This classification method would lack the quantitative physical mechanism, which may result in the criterion of mode based primarily on subjective experience.

The mechanism of drop breakup is a key and hot research area of secondary atomization. The investigation [67] shows the structure and location of turbulent eddies, which do not appear to correlate with drop breakup morphology. The average gas flow fields show no significant differences of drop morphology between bag breakup and sheet-thinning breakup. The results show that the wake structure of gas is unlikely to be the dominant mechanism of secondary atomization. These results agree well with experimental photos that the morphological transition of drop breakup is a strong function of We, and the influence of Re_g is little [68].

Interfacial instability is very important in atomization [69, 70]. Rayleigh-Taylor (RT) instability is considered as the main driving mechanism responsible for drop breakup in the general bag breakup or Rayleigh-Taylor piercing (including bag breakup, bag-stamen breakup, dual-bag breakup, bag/plume breakup, multibag breakup, etc.) [65, 71–77]. All of these breakup modes have the same characteristic bag structure. With the increase of Weber number, the thin sheet (or membrane) at the periphery of deforming drop appears and breaks up continuously. These breakup types that have continuing shearing and entraining action are all governed by the Kelvin-Helmholtz (KH) instability mainly. This mode can be named as shear breakup (or sheet-thinning breakup, shear-induced entrainment) [65, 78]. So based on the instability in secondary atomization, there are two main modes: general bag breakup and shear breakup.

In order to classify the submode of general bag breakup, a dimensionless number of RT instability wave number is proposed [71]

$$N_{RT} = D/\lambda_{RT}, \tag{17}$$

where λ_{RT} is the wavelength of the most R-T unstable wave. N_{RT} is the number of R-T wave on the windward side, which can also be considered as the bag number

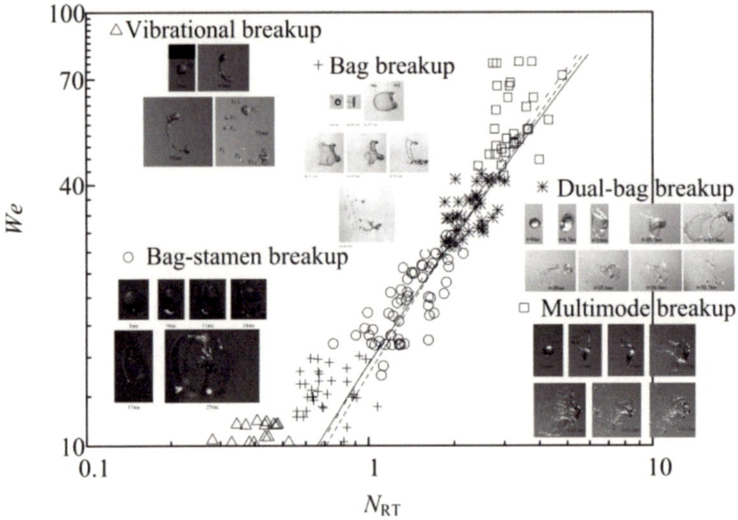

Figure 3.
Theoretical criterion N_{RT} for general bag breakup.

approximately. So the theoretical criterion N_{RT} could be the new criterion instead of *We* in the range of general bag breakup as shown in **Figure 3**.

When the viscosity of liquid cannot be neglected, *Oh* will be another key parameter [79–84]. Many researches show that the *We* range of drop breakup mode will increase with the increase of *Oh* nonlinearly. The most important transition *We* is the critical Weber number We_c occurring at the start of bag breakup. It can establish the criteria for the onset of secondary atomization. Based on the RT instability, the theoretical formula for predicting We_c is [81]

$$\left(\frac{We_0}{We_c}\right)^{1/2} + C\left(\frac{Oh^2}{We_c}\right)^{1/3} = 1, \tag{18}$$

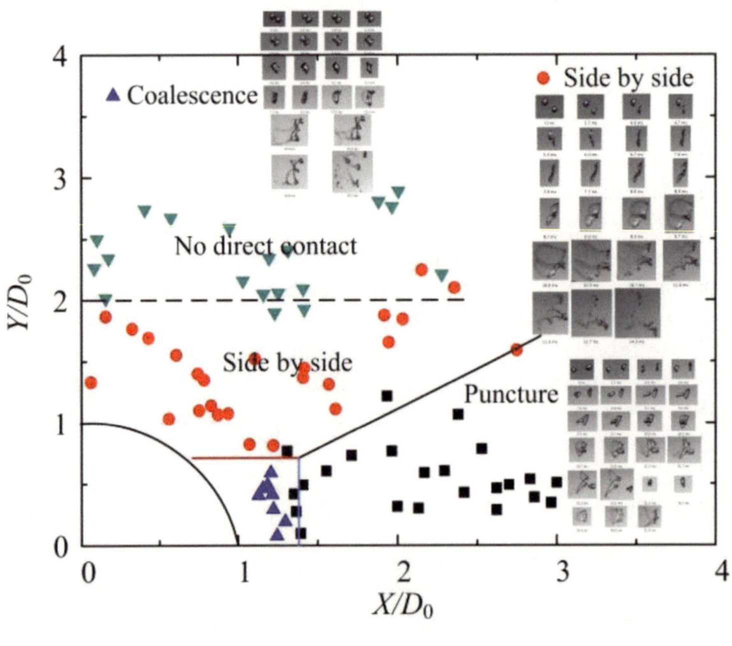

Figure 4.
Drop interaction regime map.

where We_0 is the critical Weber number when Ohnesorge number tends to zero, $We_0 = 11 \pm 2$, and C can be considered as a constant 1.48.

Drop interaction in the continuous gas jet is the important atomization process between primary atomization and secondary atomization [85]. It is the important link between the jet breakup and final spray. Due to airflow, the interaction of two drops evolves in a highly interactive and variable manner. The drop interaction in the airflow yields even more rich atomization morphologies and mechanisms. Behaviors of drop group and the isolated drop in the airflow are significantly different. Experimental photos in **Figure 4** show that there are four main interaction modes, which are coalescence mode, puncture mode, side by side mode, and no direct contact mode [86].

5. Fragment size and distribution

Drop size in atomization is a key parameter that is needed for a lot of fundamental researches and applications [87]. Due to the complicated nature of atomization, most nozzles cannot produce sprays of uniform droplet size. Instead, the spray can be regarded as a spectrum of drop sizes distributed about some defined mean drop size. Now, the most widely used mean diameter is Sauter mean diameter,

$$D_{32} \text{ or } SMD = \frac{\sum n_i d_i^3}{\sum n_i d_i^2} \tag{19}$$

where n_i is the number of droplets per unit volume in size class i, and d_i is the droplet diameter [7, 88–91].

The liquid in prefilming air-blast nozzle is first spread into a very thin sheet or film, which is then exposed to gas operating at the high velocity causing breakup and atomization. By spreading bulk liquid into film, contact area between liquid and gas increases. Generally speaking, SMD will decrease with the increase of gas velocity. However, under some conditions of prefilming atomization, the droplet size increases with the increase of gas velocity, and then decreases with the increase of gas velocity. So, the classical KH-RT atomization model [92–94] is modified and extended to the prefilming air-blast atomization [95].

Droplet size distribution is a crucial parameter of atomization process besides droplet mean diameter. Atomization and spray presents a wide distribution of fragment sizes. Many empirical relationships have been proposed to characterize the distribution of droplet sizes in atomization, for example, Rosin-Rammler, Nukiyama-Tanasawa, log-normal, root-normal, and log-hyperbolic. Atomization process involves a succession of changes of liquid topology, the last being the elongation and capillary breakup of ligaments torn off from the liquid surface. Breakup of liquid ligament (filament or fiber) is the key in primary atomization and secondary atomization, so ligament-mediated spray formation model is proposed [96, 97]. Drop fragments after ligament breakup is found to be gamma distribution. Then, the broad statistics of atomization shows Marshall-Palmer exponential shape of overall distribution in spray [98–101].

6. Complex fluids

Complex fluids are mixtures that have a coexistence between two or multi-phases, which are common in our society and industry [102–104]. Many complex fluids are non-Newtonian fluid, whose characteristics of breakup and atomization are unusual [82, 105–107].

The particle concentration in the pinch-off zone of suspension or slurry decreases as its minimal diameter decreases, resulting in a pure liquid interstitial fluid. There are three successive stages during suspension pinch-off, referred to as suspension, transition, and liquid stages, which is different from pure liquids [108–111].

For evaluating the breakup of non-Newtonian fluid, the mean apparent viscosity of liquid during deformation and breakage is the key parameter. Three methods for determining the apparent viscosity of non-Newtonian fluid have been presented: (1) calculation of mean apparent viscosity according to the shear rate equal to $\gamma = u_g/D$ [112], (2) increase the constant k determined by other test parameters, $\gamma = ku_g/D$ [17, 113], and (3) numerical analysis or analytical solution of energy and motion equations to determine dynamic shear rate [114–117].

Based on morphology, the breakup regimes of slurry jet can be classified into different modes: Rayleigh-type breakup, fiber-type breakup, superpulsating breakup, and atomization [113, 118, 119]. The particles in slurry will make membrane breakup very fast, so the membrane structure is not obvious in slurry atomization as shown in **Figure 5**. The dimensionless slurry jet breakup length can be correlated by the KH-RT hybrid model [92, 93, 113, 120]. There are two kinds of periodic structures, which are shear wave and jet oscillation. The deformation and breakup regimes of slurry drops can be classified into different modes: deformation, multimode breakup (including two submodes: hole breakup and tensile breakup), and shear breakup [17, 117].

Atomization of solution is a common phenomenon in numerous practical applications [121–124]. In the breakup of surfactant-laden liquid, critical micelle concentration (CMC) has an important influence [125–129]. The micelle can be considered as the source term, which can supply the monomers [130]. The diffusion rate of surfactant is limited, and liquid breakup is very fast sometimes. So, dynamic surface tension will change with the process of liquid deformation and atomization. According to the competition of the amplification rate of KH instability ω_{km} and RT instability ω_{Rm}, the criterion on transition Weber number between general bag breakup and shear breakup is obtained [131],

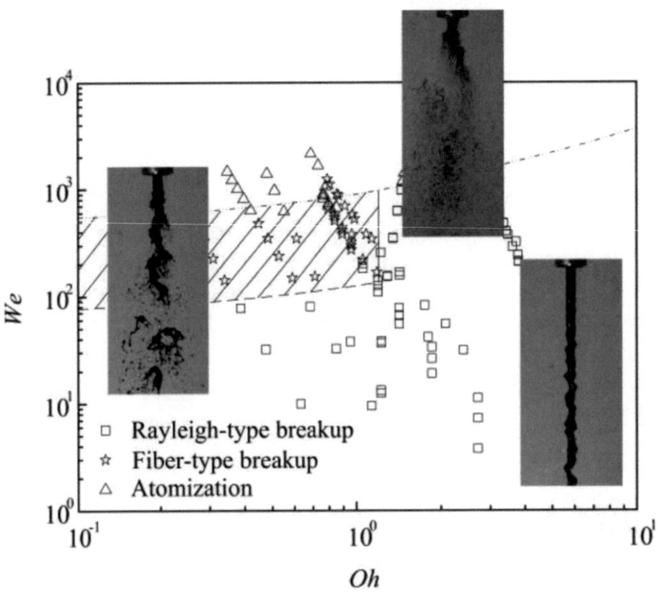

Figure 5.
The breakup regimes of slurry jet.

$$\omega_{km}/\omega_{Rm} = 1. \tag{20}$$

Atomization of viscoelastic liquids is widely known to be more difficult to atomize than typical Newtonian liquids [18, 132–138]. The addition of viscoelasticity is found to stabilize the rim of liquid. Viscoelasticity can enhance the growth of the bead and delay pinch off. Viscoelasticity increases the mean drop diameter and broadens the size distribution. Liquids with atypical properties, such as gels, liquid metal, and strain-thickening liquids, are also studied widely [139–147].

7. Conclusions

The available literature on liquid atomization is countless. Many researchers and engineers have done a lot of excellent work in this field. Unfortunately, the clear physical mechanisms on atomization have not yet been fully revealed. Some topics have received only cursory attention, such as non-Newtonian liquids, charged liquids, and turbulence influence. There are many challenges ahead for research in atomization and spray technology [148, 149]. On the other hand, it is lucky for us. Due to the fundamental nature of the problem and its many important applications, we can expect great progress in the fields of atomization and spray technology in the future.

Acknowledgements

This work was supported by the National Key R&D Program of China (2018YFC0808502-02), National Natural Science Foundation of China (21506059), Shanghai Engineering Research Center of Coal Gasification (18DZ2283900), and Fundamental Research Funds for the Central Universities.

Conflict of interest

The authors have declared that no conflict of interest exists.

Author details

Hui Zhao and Haifeng Liu*
Key Laboratory of Coal Gasification and Energy Chemical Engineering of Ministry of Education, Shanghai Engineering Research Center of Coal Gasification, East China University of Science and Technology, Shanghai, People's Republic of China

*Address all correspondence to: hfliu@ecust.edu.cn

References

[1] Eggers J, Villermaux E. Physics of liquid jets. Reports on Progress in Physics. 2008;**71**(3):036601. DOI: 10.1088/0034-4885/71/3/036601

[2] Villermaux E. Fragmentation. Annual Review of Fluid Mechanics. 2007;**39**(1):419-446. DOI: 10.1146/annurev.fluid.39.050905.110214

[3] Gorokhovski M, Herrmann M. Modeling primary atomization. Annual Review of Fluid Mechanics. 2008;**40**(1):343-366. DOI: 10.1146/annurev.fluid.40.111406.102200

[4] Theofanous TG. Aerobreakup of Newtonian and viscoelastic liquids. Annual Review of Fluid Mechanics. 2011;**43**(1):661-690. DOI: 10.1146/annurev-fluid-122109-160638

[5] Guildenbecher DR, López-Rivera C, Sojka PE. Secondary atomization. Experiments in Fluids. 2009;**46**(3):371. DOI: 10.1007/s00348-008-0593-2

[6] Dumouchel C. On the experimental investigation on primary atomization of liquid streams. Experiments in Fluids. 2008;**45**(3):371-422. DOI: 10.1007/s00348-008-0526-0

[7] Suh HK, Chang SL. A review on atomization and exhaust emissions of a biodiesel-fueled compression ignition engine. Renewable & Sustainable Energy Reviews. 2016;**58**:1601-1620. DOI: 10.1016/j.rser.2015.12.329

[8] Qian L, Lin J. Modeling on effervescent atomization: A review. Science China - Physics Mechanics & Astronomy. 2011;**54**(12):2109-2129. DOI: 10.1007/s11433-011-4536-1

[9] Boggavarapu P, Ravikrishna RV. A review on atomization and sprays of biofuels for IC engine applications. International Journal of Spray and Combustion Dynamics. 2013;**5**(2):85-122. DOI: 10.1260/1756-8277.5.2.85

[10] Basaran OA, Gao H, Bhat PP. Nonstandard inkjets. Annual Review of Fluid Mechanics. 2013;**45**:85-113. DOI: 10.1146/annurev-fluid-120710-101148

[11] Basu S, Agarwal AK, Mukhopadhyay A, Patel C. Droplets and Sprays, Applications for Combustion and Propulsion. Singapore: Springer; 2018. 430p. DOI: 10.1007/978-981-10-7449-3

[12] Sazhin S. Droplets and Sprays. London: Springer; 2014. 345p. DOI: 10.1007/978-981-10-7449-3

[13] Ashgriz N. Handbook of Atomization and Sprays. New York: Springer; 2011. 935p. DOI: 10.1007/978-1-4419-7264-4

[14] Lefebvre AH, McDonell VG. Atomization and Sprays, Second Edition. Boca Raton: CRC Press; 2017. 284p. ISBN: 9781498736268

[15] Nasr GG, Yule AJ, Bendig L. Industrial Sprays and Atomization Design, Analysis and Applications. Berlin: Springer; 2002. 501p. DOI: 10.1007/978-1-4471-3816-7

[16] Nicholls JA, Ranger AA. Aerodynamic shattering of liquid drops. AIAA Journal. 1968;**7**(2):285-290. DOI: 10.2514/6.1968-83

[17] Zhao H, Liu HF, Xu JL, Li WF. Secondary breakup of coal water slurry drops. Physics of Fluids. 2011;**23**(11):074501. DOI: 10.1063/1.3659495

[18] Zhao H, Hou YB, Liu HF, Tian XS, Xu JL, Li WF, et al. Influence of rheological properties on air-blast atomization of coal water slurry. Journal of Non-Newtonian Fluid Mechanics.

2014;**211**:1-15. DOI: 10.1016/j.jnnfm.2014.06.007

[19] Zhong W, He Z, Wang Q, Shao Z, Tao X. Experimental study of flow regime characteristics in diesel multi-hole nozzles with different structures and enlarged scales. International Communications in Heat and Mass Transfer. 2014;**59**:1-10. DOI: 10.1016/j.icheatmasstransfer.2014.10.001

[20] Lahane S, Subramanian KA. Impact of nozzle holes configuration on fuel spray, wall impingement and NOx emission of a diesel engine for biodiesel-diesel blend (B20). Applied Thermal Engineering. 2014;**64**(1–2):307-314. DOI: 10.1016/j.applthermaleng.2013.12.048

[21] Taghavifar H, Shervanitabar MT, Abbasalizadeh M. Numerical study of the effects of injector needle movement and the nozzle inclination angle on the internal fluid flow and spray structure of a group-hole nozzle layout. Applied Mathematical Modelling. 2015;**39**(23–24):7718-7733. DOI: 10.1016/j.apm.2015.04.032

[22] Xue Q, Battistoni M, Powell CF, Longman DE, Quan SP, Pomraning E, et al. An Eulerian CFD model and X-ray radiography for coupled nozzle flow and spray in internal combustion engines. International Journal of Multiphase Flow. 2015;**70**:77-88. DOI: 10.1016/j.ijmultiphaseflow.2014.11.012

[23] Fansler TD, Parrish SE. Spray measurement technology: A review. Measurement Science & Technology. 2014;**26**(1):012002. DOI: 10.1088/0957-0233/26/1/012002

[24] Gao J, Li W, Wong SP, Hu MJ, Li RKY. Controllable morphology and wettability of polymer microspheres prepared by nonsolvent assisted electrospraying. Polymer. 2014;**55**(12): 2913-2920. DOI: 10.1016/j.polymer.2014.04.033

[25] Chandrasekhar PS, Kumar N, Swami SK, Dutta V, Komarala VK. Fabrication of perovskite films using an electrostatic assisted spray technique: The effect of the electric field on morphology, crystallinity and solar cell performance. Nanoscale. 2016;**8**(12): 6792-6800. DOI: 10.1039/c5nr08350h

[26] Liparoti S, Adami R, Reverchon E. Supercritical assisted atomization: Effect of operative conditions on PVP microparticle size and morphology. The Journal of Supercritical Fluids. 2015;**97**: 31-35. DOI: 10.1016/j.supflu.2014.10.028

[27] Jafari-Nodoushan M, Barzin J, Mobedi H. Size and morphology controlling of PLGA microparticles produced by electro hydrodynamic atomization. Polymers for Advanced Technologies. 2015;**26**(5):502-513. DOI: 10.1002/pat.3480

[28] Stunda-Zujeva A, Irbe Z, Berzina-Cimdina L. Controlling the morphology of ceramic and composite powders obtained via spray drying—A review. Ceramics International. 2017;**43**: 11543-11551. DOI: 10.1016/j.ceramint.2017.05.023

[29] Schumaker SA, Driscoll JF. Mixing properties of coaxial jets with large velocity ratios and large inverse density ratios. Physics of Fluids. 2012;**24**(5): 055101. DOI: 10.1063/1.4711396

[30] Jarrahbashi D, Sirignano WA, Popov PP, Hussain F. Early spray development at high gas density: Hole, ligament and bridge formations. Journal of Fluid Mechanics. 2016;**792**:186-231. DOI: 10.1017/jfm.2016.71

[31] Padwal MB, Jejurkar SY, Mishra DP. Experimental studies on bluff body-assisted airblast atomizer. Atomization and Sprays. 2016;**26**(11):1127-1150. DOI: 10.1615/AtomizSpr.2015013523

[32] Wahono S, Honnery D, Soria J, Ghojel J. High-speed visualisation of

primary break-up of an annular liquid sheet. Experiments in Fluids. 2008;**44**: 451-459. DOI: 10.1007/s00348-007-0361-8

[33] Duke D, Honnery D, Soria J. A cross-correlation velocimetry technique for breakup of an annular liquid sheet. Experiments in Fluids. 2010;**49**: 435-445. DOI: 10.1007/s00348-009-0817-0

[34] Duke D, Honnery D, Soria J. Experimental investigation of nonlinear instabilities in annular liquid sheets. Journal of Fluid Mechanics. 2012;**691**: 594-604. DOI: 10.1017/jfm.2011.516

[35] Gañán-Calvo AM, Herrada MA, Garstecki P. Bubbling in unbounded coflowing liquids. Physical Review Letters. 2006;**96**:124504. DOI: 10.1103/PhysLettRev.96.124504

[36] Farago Z, Chigier N. Morphological classification of disintegration of round liquid jets in a coaxial air stream. Atomization and Sprays. 1992;**2**:137-153. DOI: 10.1615/AtomizSpr.v2.i2.50

[37] Lasheras JC, Villermaux E, Hopfinger EJ. Break-up and atomization of a round water jet by a high-speed annular air jet. Journal of Fluid Mechanics. 1998;**357**:351-379. DOI: 10.1017/S0022112097008070

[38] Lasheras JC, Hopfinger EJ. Liquid jet instability and atomization in a coaxial gas stream. Annual Review of Fluid Mechanics. 2000;**32**(1):275-308. DOI: 10.1146/annurev.fluid.32.1.275

[39] Wang Y, Im KS, Fezzaa K. Similarity between the primary and secondary air-assisted liquid jet breakup mechanisms. Physical Review Letters. 2008;**100**(15):154502. DOI: 10.1103/PhysRevLett.100.154502

[40] Baillot F, Blaisot JB, Boisdron G, Dumouchel C. Behaviour of an air-assisted jet submitted to a transverse high-frequency acoustic field. Journal of Fluid Mechanics. 2009;**640**(640): 305-342. DOI: 10.1017/S002211200999139X

[41] Zhao H, Liu HF, Tian XS, Xu JL, Li WF, Lin KF. Influence of atomizer exit area ratio on the breakup morphology of coaxial air and round water jets. AIChE Journal. 2014;**60**(6):2335-2345. DOI: 10.1002/aic.14414

[42] Kumar A, Sahu S. Liquid jet breakup unsteadiness in a coaxial air-blast atomizer. International Journal of Spray and Combustion Dynamics. 2018;**10**(3): 211-230. DOI: 10.1177/1756827718760905

[43] Kendall JM. Experiments on annular liquid jet instability and on the formation of liquid shells. Physics of Fluids. 1986;**29**: 2086-2094. DOI: 10.1063/1.865595

[44] Lee JG, Chen LD. Linear stability analysis of gas-liquid interface. AIAA Journal. 1991;**29**(10):1589-1595. DOI: 10.2514/3.10779

[45] Martínez-Bazán C, Montañés JL, Lasheras JC. On the breakup of an air bubble injected into a fully developed turbulent flow. Part 1. Breakup frequency. Journal of Fluid Mechanics. 1999;**401**:157-182. DOI: 10.1017/S0022112099006680

[46] Martínez-Bazán C, Montañés JL, Lasheras JC. On the breakup of an air bubble injected into a fully developed turbulent flow. Part 2. Size PDF of the resulting daughter bubbles. Journal of Fluid Mechanics. 1999;**401**:183-207. DOI: 10.1017/S0022112099006692

[47] Adzic M, Carvalho IS, Heitor MV. Visualisation of the disintegration of an annular liquid sheet in a coaxial airblast injector at low atomising air velocities. Optical Diagnostics in Engineering. 2001;**5**:27-38

[48] Li XG, Shen JH. Experiments on annular liquid jet breakup. Atomization

and Sprays. 2001;**11**:557-573. DOI: 10.1615/AtomizSpr.v11.i5.50

[49] Choi CJ, Lee SY. Droplet formation from thin hollow liquid jet with a core air flow. Atomization and Sprays. 2005; **15**:469-487. DOI: 10.1615/AtomizSpr.v15.i4.80

[50] Sevilla A, Gordillo JM, Martínez-Bazán C. Transition from bubbling to jetting in a coaxial air-water jet. Physics of Fluids. 2005;**17**:018105. DOI: 10.1063/1.1831312

[51] Sevilla A, Gordillo JM, Martínez-Bazán C. Bubble formation in a coflowing air-water stream. Journal of Fluid Mechanics. 2005;**530**:181-195. DOI: 10.1017/S002211200500354X

[52] Leboucher N, Roger F, Carreau JL. Atomization characteristics of an annular liquid sheet with inner and outer gas flows. Atomization and Sprays. 2014;**24**(12):1065-1088. DOI: 10.1615/AtomizSpr.2014010670

[53] Zhao H, Liu HF, Tian XS, Xu JL, Li WF, Lin KF. Outer ligament-mediated spray formation of annular liquid sheet by an inner round air stream. Experiments in Fluids. 2014;**55**(8): 1793-1780. DOI: 10.1007/s00348-014-1793-6

[54] Zhao H, Xu JL, Wu JH, Li WF, Liu HF. Breakup morphology of annular liquid sheet with an inner round air stream. Chemical Engineering Science. 2015;**137**:412-422. DOI: 10.1016/j.ces.2015.06.062

[55] Fu QF, Deng XD, Jia BQ, Yang LJ. Temporal instability of a confined liquid film with heat and mass transfer. AIAA Journal. 2018:1-8. DOI: 10.2514/1.J056834

[56] Schefer R, Namazian M, Kelly J. Velocity measurements in turbulent bluff-body stabilized flows. AIAA Journal. 1994;**32**:1844-1851. DOI: 10.2514/3.12182

[57] Yang JT, Chang CC, Pan KL. Flow structures and mixing mechanisms behind a disc stabilizer with a central fuel jet. Combustion Science and Technology. 2002;**174**(3):93-124. DOI: 10.1080/00102200290021001

[58] Riber E, Moureau V, García M, Poinsot T, Simonin O. Evaluation of numerical strategies for large eddy simulation of particulate two-phase recirculating flows. Journal of Computational Physics. 2009;**228**(2): 539-564. DOI: 10.1016/j.jcp.2008.10.001

[59] Tian XS, Zhao H, Liu HF, Li WF, Xu JL. Liquid entrainment behavior at the nozzle exit in coaxial gas-liquid jets. Chemical Engineering Science. 2014; **107**:93-101. DOI: 10.1016/j.ces.2013.12.008

[60] Tian XS, Zhao H, Liu HF, Li WF, Xu JL. Effect of central tube thickness on wave frequency of coaxial liquid jet. Fuel Processing Technology. 2014;**119**: 190-197. DOI: 10.1016/j.fuproc.2013.11.011

[61] Gujeong P, Jungho L, Ingyu L, Yoon Y, Hoon SC. Geometric effect on spray characteristics of gas-centered swirl coaxial injectors: Recess ratio and gap thickness. Atomization and Sprays. 2017;**27**(7):579-589. DOI: 10.1615/AtomizSpr.2017018958

[62] Chou WH, Hsiang LP, Faeth GM. Temporal properties of drop breakup in the shear breakup regime. International Journal of Multiphase Flow. 1997;**23**(4): 651-669. DOI: 10.1016/S0301-9322(97)00006-2

[63] Chou WH, Faeth GM. Temporal properties of secondary drop breakup in the bag breakup regime. International Journal of Multiphase Flow. 1998;**24**(6): 889-912. DOI: 10.1016/S0301-9322(98)00015-9

[64] Dai Z, Faeth GM. Temporal properties of secondary drop breakup in the multimode breakup regime. International Journal of Multiphase Flow. 2001;**27**(2):217-236. DOI: 10.1016/S0301-9322(00)00015-X

[65] Theofanous TG, Li GJ, Dinh TN. Aerobreakup in rarefied supersonic gas flows. Journal of Fluids Engineering. 2004;**126**(4):516-527. DOI: 10.1115/1.1777234

[66] Zhao H, Liu HF, Xu JL, Li WF, Lin KF. Temporal properties of secondary drop breakup in the bag-stamen breakup regime. Physics of Fluids. 2013;**25**(5):1741. DOI: 10.1063/1.4803154

[67] Flock AK, Guildenbecher DRR, Chen J, Sojka PE, Bauer H-J. Experimental statistics of droplet trajectory and air flow during aerodynamic fragmentation of liquid drops. International Journal of Multiphase Flow. 2012;**47**(3):37-49. DOI: 10.1016/j.ijmultiphaseflow.2012.06.008

[68] Lee CH, Reitz RD. An experimental study of the effect of gas density on the distortion and breakup mechanism of drops in high speed gas stream. International Journal of Multiphase Flow. 2000;**26**(2):229-244. DOI: 10.1016/S0301-9322(99)00020-8

[69] Reitz R. Modeling atomization processes in high-pressure vaporizing sprays. Atomization and Spray Technology. 1987;**3**(4):309-337

[70] Joseph DD, Belanger J, Beavers GS. Breakup of a liquid drop suddenly exposed to a high-speed airstream. International Journal of Multiphase Flow. 1999;**25**(6–7):1263-1303. DOI: 10.1016/S0301-9322(99)00043-9

[71] Zhao H, Liu HF, Li WF, Xu JL. Morphological classification of low viscosity drop bag breakup in a continuous air jet stream. Physics of Fluids. 2010;**22**(11):114103. DOI: 10.1063/1.3490408

[72] Rimbert N, Castanet G. Crossover between Rayleigh-Taylor instability and turbulent cascading atomization mechanism in the bag-breakup regime. Physical Review. E, Statistical, Nonlinear, and Soft Matter Physics. 2011;**84**(1 Pt 2):016318. DOI: 10.1103/PhysRevE.84.016318

[73] Jalaal M, Mehravaran K. Transient growth of droplet instabilities in a stream. Physics of Fluids. 2014;**26**(1):012101. DOI: 10.1063/1.4851056

[74] Jain M, Prakash RS, Tomar G, Ravikrishna RV. Secondary breakup of a drop at moderate Weber numbers. Proceedings of the Royal Society of London. 2015;**471**(2177):20140930. DOI: 10.1098/rspa.2014.0930

[75] Yang W, Jia M, Che Z, Sun K, Wang TY. Transitions of deformation to bag breakup and bag to bag-stamen breakup for droplets subjected to a continuous gas flow. International Journal of Heat and Mass Transfer. 2017;**111**:884-894. DOI: 10.1016/j.ijheatmasstransfer.2017.04.012

[76] Xiao F, Dianat M, Mcguirk JJ. A robust interface method for drop formation and breakup simulation at high density ratio using an extrapolated liquid velocity. Computers & Fluids. 2016;**136**:402-420. DOI: 10.1016/j.compfluid.2016.06.021

[77] Xiao F, Wang ZG, Sun MB, Liu N, Yang X. Simulation of drop deformation and breakup in supersonic flow. Proceedings of the Combustion Institute. 2017;**36**(2):2417-2424. DOI: 10.1016/j.proci.2016.09.016

[78] Theofanous TG, Li GJ. On the physics of aerobreakup. Physics of Fluids. 2008;**20**(5):052103. DOI: 10.1063/1.2907989

[79] Hsiang LP, Faeth GM. Drop deformation and breakup due to shock wave and steady disturbances. International Journal of Multiphase Flow. 1994;**21**(4):545-560. DOI: 10.1016/0301-9322(94)00095-2

[80] Zhao H, Liu HF, Cao XK, Li WF, Xu JL. Breakup characteristics of liquid drops in bag regime by a continuous and uniform air jet flow. International Journal of Multiphase Flow. 2011;**37**(5): 530-534. DOI: 10.1016/j.ijmultiphaseflow.2010.12.006

[81] Theofanous TG, Mitkin VV, Ng CL, Chang C-H, Deng X, Sushchikh S. The physics of aerobreakup. II. Viscous liquids. Physics of Fluids. 2012;**24**(2):022104. DOI: 10.1063/1.3680867

[82] Müller T, Sänger A, Habisreuther P, Jakobs T, Trimis D, Kolb T, et al. Simulation of the primary breakup of a high-viscosity liquid jet by a coaxial annular gas flow. International Journal of Multiphase Flow. 2016;**87**:212-228. DOI: 10.1016/j.ijmultiphaseflow.2016.09.008

[83] Stefanitsis D, Malgarinos I, Strotos G, Nikolopoulos N, Kakaras E, Gavaises M. Numerical investigation of the aerodynamic breakup of diesel and heavy fuel oil droplets. International Journal of Heat and Fluid Flow. 2017;**68**: 203-215. DOI: 10.1016/j.ijheatfluidflow.2017.10.012

[84] Strotos G, Malgarinos I, Nikolopoulos N, Gavaises M, Nikas KS, Moustris K. Determination of the aerodynamic droplet breakup boundaries based on a total force approach. International Journal of Heat and Fluid Flow. 2018;**69**:164-173. DOI: 10.1016/j.ijheatfluidflow.2018.01.001

[85] Theofanous TG, Li GJ, Dinh TN, Chang CH. Aerobreakup in disturbed subsonic and supersonic flow fields. Journal of Fluid Mechanics. 2007; **593**(593):131-170. DOI: 10.1017/S0022112007008853

[86] Zhao H, Wu ZW, Li WF, Xu JL, Liu HF. Interaction of two drops in the bag breakup regime by a continuous air jet. Fuel. 2019;**236**:843-850. DOI: 10.1016/j.fuel.2018.09.067

[87] Babinsky E, Sojka PE. Modeling drop size distributions. Progress in Energy and Combustion Science. 2002; **28**(4):303-329. DOI: 10.1016/S0360-1285(02)00004-7

[88] Lee S, Park S. Experimental study on spray break-up and atomization processes from GDI injector using high injection pressure up to 30MPa. International Journal of Heat and Fluid Flow. 2014;**45**:14-22. DOI: 10.1016/j.ijheatfluidflow.2013.11.005

[89] Ghasemi A, Barron RM, Balachandar R. Spray-induced air motion in single and twin ultra-high injection diesel sprays. Fuel. 2014; **121**(2):284-297. DOI: 10.1016/j.fuel.2013.12.041

[90] Agarwal AK, Dhar A, Gupta JG, Kim WI, Choi K, Lee CS, et al. Effect of fuel injection pressure and injection timing of Karanja biodiesel blends on fuel spray, engine performance, emissions and combustion characteristics. Energy Conversion and Management. 2015;**91**:302-314. DOI: 10.1016/j.enconman.2014.12.004

[91] Mo J, Tang C, Li J, Guan L, Huang ZH. Experimental investigation on the effect of n-butanol blending on spray characteristics of soybean biodiesel in a common-rail fuel injection system. Fuel. 2016;**182**:391-401. DOI: 10.1016/j.fuel.2016.05.109

[92] Varga CM, Lasheras JC, Hopfinger EJ. Initial breakup of a small-diameter liquid jet by a high-speed gas stream. Journal of Fluid Mechanics.

2003;**497**:405-434. DOI: 10.1017/S0022112003006724

[93] Aliseda A, Hopfinger EJ, Lasheras JC, Kremer DM, Berchielli A, Connolly EK. Atomization of viscous and non-Newtonian liquids by a coaxial, high-speed gas jet. Experiments and droplet size modeling. International Journal of Multiphase Flow. 2008;**34**(2):161-175. DOI: 10.1016/j.ijmultiphaseflow.2007.09.003

[94] Vadivukkarasan M, Panchagnula MV. Combined Rayleigh-Taylor and Kelvin-Helmholtz instabilities on an annular liquid sheet. Journal of Fluid Mechanics. 2017;**812**:152-177. DOI: 10.1017/jfm.2016.784

[95] Zhao H, Wu ZW, Li WF, Xu JL, Liu HF. Nonmonotonic effects of aerodynamic force on droplet size of prefilming air-blast atomization. Industrial & Engineering Chemistry Research. 2018;**57**(5):1726-1732. DOI: 10.1021/acs.iecr.7b05026

[96] Villermaux E, Marmottant P, Duplat J. Ligament-mediated spray formation. Physical Review Letters. 2004;**92**(7):074501. DOI: 10.1103/PhysRevLett.92.074501

[97] Marmottant P, Villermaux E. On spray formation. Journal of Fluid Mechanics. 2004;**498**(498):73-111. DOI: 10.1017/S0022112003006529

[98] Villermaux E, Bossa B. Single-drop fragmentation determines size distribution of raindrops. Nature Physics. 2009;**5**(9):697-702. DOI: 10.1038/NPHYS1340

[99] Villermaux E, Bossa B. Drop fragmentation on impact. Journal of Fluid Mechanics. 2011;**668**(4):412-435. DOI: 10.1017/S002211201000474X

[100] Zhao H, Liu HF, Xu JL, Li WF. Experimental study of drop size distribution in the bag breakup regime. Industrial & Engineering Chemistry Research. 2011;**50**(16):9767-9773. DOI: 10.1021/ie200622d

[101] Stefan K, Rick S, Denn MM, Villermaux E, Bonn D. What determines the drop size in sprays. Physical Review X. 2018;**8**(3):031019. DOI: 10.1103/PhysRevX.8.031019

[102] Divoux T, Fardin MA, Manneville S, Lerouge S. Shear banding of complex fluids. Annual Review of Fluid Mechanics. 2015;**48**(48):81-103. DOI: 10.1146/annurev-fluid-122414-034416

[103] Waigh TA. Advances in the microrheology of complex fluids. Reports on Progress in Physics. 2016;**79**(7):074601. DOI: 10.1088/0034-4885/79/7/074601

[104] Patteson AE, Gopinath A, Arratia PE. Active colloids in complex fluids. Current Opinion in Colloid & Interface Science. 2016;**21**:86-96. DOI: 10.1016/j.cocis.2016.01.001

[105] Keshavarz B, Sharma V, Houze EC, Koerner MR, Moore JR, Cotts PM, et al. Studying the effects of elongational properties on atomization of weakly viscoelastic solutions using Rayleigh Ohnesorge Jetting Extensional Rheometry (ROJER). Journal of Non-Newtonian Fluid Mechanics. 2015;**222**:171-189. DOI: 10.1016/j.jnnfm.2014.11.004

[106] Broniarz-Press L, Sosnowski TR, Matuszak M, Ochowiak M, Jablczynska K. The effect of shear and extensional viscosities on atomization of Newtonian and non-Newtonian fluids in ultrasonic inhaler. International Journal of Pharmaceutics. 2015;**485**(1–2):41-49. DOI: 10.1016/j.ijpharm.2015.02.065

[107] Bienia M, Lejeune M, Chambon M, Baco-Carles V, Dossou-Yovo C, Noguera R, et al. Inkjet printing of ceramic colloidal suspensions: Filament growth and breakup. Chemical

Engineering Science. 2016;**149**:1-13. DOI: 10.1016/j.ces.2016.04.015

[108] Zhao H, Liu HF, Xu JL, Li WF, Lin KF. Inhomogeneity in breakup of suspensions. Physics of Fluids. 2015; **27**(6):419. DOI: 10.1063/1.4922582

[109] Mathues W, Mcilroy C, Harlen OG, Clasen C. Capillary breakup of suspensions near pinch-off. Physics of Fluids. 2015;**27**(9):093301. DOI: 10.1063/1.4930011

[110] Zou J, Lin F, Ji C. Capillary breakup of armored liquid filaments. Physics of Fluids. 2017;**29**(6):062103. DOI: 10.1063/1.4984836

[111] Mariano RR, Wouter M, Alejandro S, Clasen C. One-dimensional modelling of the thinning of particulate suspensions near pinch-off. International Journal of Multiphase Flow. 2018;**108**:202-210. DOI: 10.1016/j.ijmultiphaseflow.2018.06.007

[112] Lopez-Rivera C, Sojka PE. Secondary breakup of non-Newtonian liquid drops. In: ILASS Europ. 22nd European Conference on Liquid Atomization and Spray Dystems. Como Lake, Italy; 2008

[113] Zhao H, Liu HF, Xu JL, Li WF, Cheng W. Breakup and atomization of a round coal water slurry jet by an annular air jet. Chemical Engineering Science. 2012;**78**:63-74. DOI: 10.1016/j.ces.2012.05.007

[114] Dechelette A, Sojka PE, Wassgren CR. Non-Newtonian drops spreading on a flat surface. Journal of Fluids Engineering. 2010;**132**(10):101302. DOI: 10.1115/1.4002281

[115] An SM, Lee SY. One-dimensional model for the prediction of impact dynamics of a shear-thinning liquid drop on dry solid surfaces. Atomization and Sprays. 2012;**22**(5):371-389. DOI: 10.1615/AtomizSpr.2012005599

[116] Focke C, Bothe D. Direct numerical simulation of binary off-center collisions of shear thinning droplets at high Weber numbers. Physics of Fluids. 2012;**24**(7):073105. DOI: 10.1063/1.4737582

[117] Tavangar S, Hashemabadi SH, Saberimoghadam A. CFD simulation for secondary breakup of coal-water slurry drops using OpenFOAM. Fuel Processing Technology. 2015;**132**: 153-163. DOI: 10.1016/j.fuproc.2014.12.037

[118] Wayne S, Francine B. The effects of pulsation and retraction on non-Newtonian flows in three-stream injector atomization systems. Chemical Engineering Journal. 2017; **309**(1):532-544. DOI: 10.1016/j.cej.2016.10.046

[119] Jampolski L, Sanger A, Jakobs T, Guthausen G, Kolb T, Willenbacher N. Improving the processability of coke water slurries for entrained flow gasification. Fuel. 2016;**185**:102-111. DOI: 10.1016/j.fuel.2016.07.102

[120] Beale JC, Reitz RD. Modeling spray atomization with the Kelvin-Helmholtz/Rayleigh-Taylor hybrid model. Atomization and Sprays. 1999; **9**(6):623-650. DOI: 10.1615/AtomizSpr.v9.i6.40

[121] Broniarz-Press L, Ochowiak M, Woziwodzki S. Atomization of PEO aqueous solutions in effervescent atomizers. International Journal of Heat and Fluid Flow. 2010;**31**(4):651-658. DOI: 10.1016/j.ijheatfluidflow.2010.02.005

[122] Stabile L, Trassierra CV, Dell'Agli G, Buonanno G. Ultrafine particle generation through atomization technique: The influence of the solution. Aerosol and Air Quality Research. 2013; **13**(6):1667-1677. DOI: 10.4209/aaqr.2013.03.0085

[123] Husain O, Lau W, Edirisinghe M, Parhizkar M. Investigating the particle to fibre transition threshold during electrohydrodynamic atomization of a polymer solution. Materials Science and Engineering C. 2016;**65**:240-250. DOI: 10.1016/j.msec.2016.03.076

[124] Zhao H, Zhang WB, Xu JL, Li WF, Liu HF. Surfactant-laden drop jellyfish-breakup mode induced by the Marangoni effect. Experiments in Fluids. 2017;**58**(3):13. DOI: 10.1007/s00348-016-2296-4

[125] Edmonstone BD, Craster RV, Matar OK. Surfactant-induced fingering phenomena beyond the critical micelle concentration. Journal of Fluid Mechanics. 2006;**564**:105-138. DOI: 10.1017/S0022112006001352

[126] Rother MA, Davis RH. Buoyancy-driven breakup of an isolated drop with surfactant. Physical Review Letters. 2008;**101**(4):044501. DOI: 10.1103/PhysRevLett.101.044501

[127] Roché M, Aytouna M, Bonn D, Kellay H. Effect of surface tension variations on the pinch-off behavior of small fluid drops in the presence of surfactants. Physical Review Letters. 2009;**103**(26):264501. DOI: 10.1103/PhysRevLett.103.264501

[128] de Saint Vincent MR, Petit J, Aytouna M, Delville JP, Bonn D, Kelly H. Dynamic interfacial tension effects in the rupture of liquid necks. Journal of Fluid Mechanics. 2012;**692**:499-510. DOI: 10.1017/jfm.2011.550

[129] Karapetsas G, Bontozoglou V. The primary instability of falling films in the presence of soluble surfactants. Journal of Fluid Mechanics. 2013;**729**:123-150. DOI: 10.1017/jfm.2013.291

[130] Zhao H, Zhang WB, Xu JL, Li WF, Liu HF. Influence of surfactant on the drop bag breakup in a continuous air jet stream. Physics of Fluids. 2016;**28**(5): 054102. DOI: 10.1063/1.4947575

[131] Zhao H, Wu ZW, Li WF, Xu JL, Liu HF. Transition Weber number between surfactant-laden drop bag breakup and shear breakup of secondary atomization. Fuel. 2018;**221**:138-143. DOI: 10.1016/j.fuel.2018.02.119

[132] Christanti Y, Walker LM. Quantifying air atomization of viscoelastic fluids through fluid relaxation times. Atomization and Sprays. 2006;**16**(7):777-790. DOI: 10.1615/AtomizSpr.v16.i7.50

[133] Thompson JC, Rothstein JP. The atomization of viscoelastic fluids in flat-fan and hollow-cone spray nozzles. Journal of Non-Newtonian Fluid Mechanics. 2007;**147**(1–2):11-22. DOI: 10.1016/jjnnfm.2007.06.004

[134] Bhat PP, Appathurai S, Harris MT, Pasquali M, McKinley GH, Basaran OA. Formation of beads-on-a-string structures during break-up of viscoelastic filaments. Nature Physics. 2010;**6**(8):625-631. DOI: 10.1038/NPHYS1682

[135] Li L, Green SI, Davy MH, Eadie DT. Viscoelastic air-blast sprays in a cross-flow. Part 1: Penetration and dispersion. Atomization and Sprays. 2010;**20**(8):697-720. DOI: 10.1615/AtomizSpr.v20.i8.30

[136] Li L, Green SI, Davy MH, Eadie DT. Viscoelastic air-blast sprays in a cross-flow. Part 2: Droplet velocities. Atomization and Sprays. 2010;**20**(8): 721-735. DOI: 10.1615/AtomizSpr.v20.i8.40

[137] Ruo AC, Chen F, Chung CA, Chang MH. Three-dimensional response of unrelaxed tension to instability of viscoelastic jets. Journal of Fluid Mechanics. 2011;**1**:1-19. DOI: 10.1017/jfm.2011.255

[138] Theofanous TG, Mitkin VV, Ng CL. The physics of aerobreakup. III. Viscoelastic liquids. Physics of Fluids. 2013;**25**(3):032101. DOI: 10.1063/1.4792712

[139] Padwal MB, Mishra DP. Interactions among synthesis, rheology, and atomization of a gelled propellant. Rheologica Acta. 2016;**55**(3):177-186. DOI: 10.1007/s00397-015-0903-6

[140] Kim H, Ko T, Kim S, Yoon W. Spray characteristics of aluminized-gel fuels sprayed using pressure-swirl atomizer. Journal of Non-Newtonian Fluid Mechanics. 2017;**249**:36-47. DOI: 10.1016/j.jrinfm.2017.08.003

[141] Jejurkar SY, Yadav G, Mishra DP. Visualizations of sheet breakup of non-Newtonian gels loaded with nanoparticles. International Journal of Multiphase Flow. 2018;**100**:57-76. DOI: 10.1016/j.ijmultiphaseflow.2017.12.003

[142] Mates SP, Settles GS. A study of liquid metal atomization using close-coupled nozzles, Part 1: Gas dynamic behavior. Atomization and Sprays. 2005;**15**(1):19-40. DOI: 10.1615/AtomizSpr.v15.i1.20

[143] Mates SP, Settles GS. A study of liquid metal atomization using close-coupled nozzles, part 2: Atomization behavior. Atomization and Sprays. 2005;**15**(1):41-60. DOI: 10.1615/AtomizSpr.v15.i1.30

[144] Lin M, Zhong M, Li Y, Yuan M, Yang Y. Numerical analysis on molten droplet hydrodynamic deformation and surface waves under high pressure pulse. Annals of Nuclear Energy. 2015;**77**:133-141. DOI: 10.1016/j.anucene.2014.11.008

[145] Manickam L, Bechta S, Ma W. On the fragmentation characteristics of melt jets quenched in water. International Journal of Multiphase Flow. 2017;**91**:262-275. DOI: 10.1016/j.ijmultiphaseflow.2017.02.005

[146] Yi C, Wagner JL, Farias PA, DeMauro EP, Guildenbecher DR. Galinstan liquid metal breakup and droplet formation in a shock-induced cross-flow. International Journal of Multiphase Flow. 2018;**106**:147-163. DOI: 10.1016/j.ijmultiphaseflow.2018.05.015

[147] Mitkin VV, Theofanous TG. The physics of aerobreakup. IV. Strain-thickening liquids. Physics of Fluids. 2017;**29**(12):122101. DOI: 10.1063/1.4997009

[148] Chigier N. Challenges for future research in atomization and spray technology: Arthur Lefebvre memorial lecture. Atomization and Sprays. 2006;**16**(7):727-736. DOI: 10.1615/AtomizSpr.v16.i7.10

[149] Jiang X, Siamas GA, Jagus K, Karayiannis TG. Physical modelling and advanced simulations of gas-liquid two-phase jet flows in atomization and sprays. Progress in Energy and Combustion Science. 2010;**36**(2):131-167. DOI: 10.1016/j.pecs.2009.09.002

Chapter 5

Modeling of Air Pollution at Airports

Oleksandr Zaporozhets and Kateryna Synylo

Abstract

Although airports provide several benefits for our society, communities in the vicinity of airports are subjected to the deterioration of air quality. Currently, the basic objects of attention are NOx and ultrafine PM due to airport-related emissions. Considered environmental problems are intensified in connection with increasing air traffic, rising tensions of airports expansion and growing cities closer and closer each other, and accordingly growing public concern with air quality around the airport. Aircraft are the dominant and special source of emission and air pollution at airports in most cases under consideration. So, to evaluate the aircraft contribution in LAQ assessment of the airports accurately, it is important to take in mind few features of the aircraft during their landing-takeoff cycle (LTO), which define emission and dispersion parameters of the considered source. The complex model PolEmiCa allows the calculation of the inventory and dispersion parameters of the aircraft engine emissions during the LTO cycles of the aircraft in the airport area. But a clear quantification of aircraft emission contribution to total air pollution is the actual task for development of cost-effective strategies to improve local air quality according to the vicinity of the airport, and to meet regulatory requirements.

Keywords: aircraft engine emission, exhaust gases jet, airport air pollution, local air quality, modeling of air pollution, emission inventory

1. Introduction

Despite significant economic and social benefits the aviation brings, its activities also contribute to local air quality impact and correspondingly affect the health and quality life of people living near the airports. The number of flights has increased by 80% between 1990 and 2014 and is forecasted to grow by a further 45% between 2014 and 2035. Consequently, the future growth in the European aviation sector will be inextricably linked to its environmental sustainability [1].

During the last decade, a lot of studies have also focused on the aircraft emissions impact on local and regional air quality in the vicinity of airport [2–7]. The basic objects of attention are extremely high concentration of toxic compounds (including nitrogen oxides (NOx), particle matter (PM with various sizes: PM10, PM2.5, and ultrafine), unburned hydrocarbons (UHC), and carbon monoxide (CO)) due to airport-related emissions and their significant impact on the environment [2, 8] and health of the people living near the airport [3, 4].

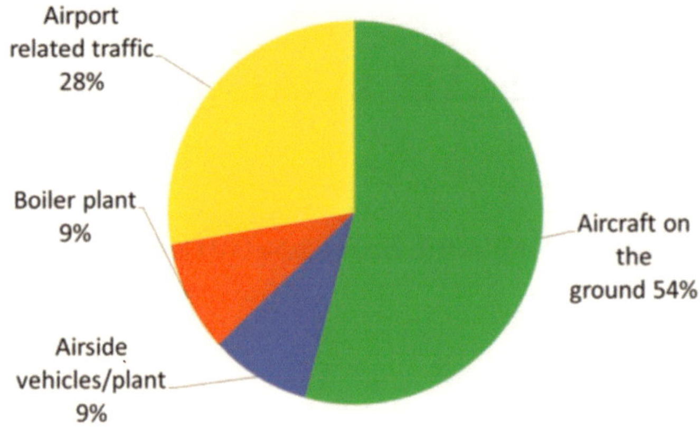

Figure 1.
Estimated ground-level airport-related emissions from Heathrow Airport.

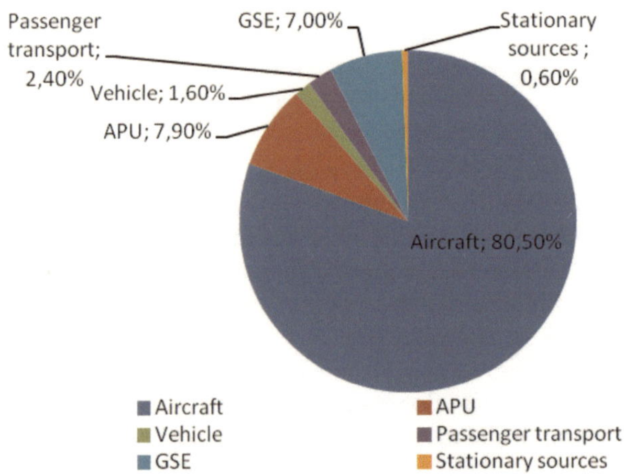

a

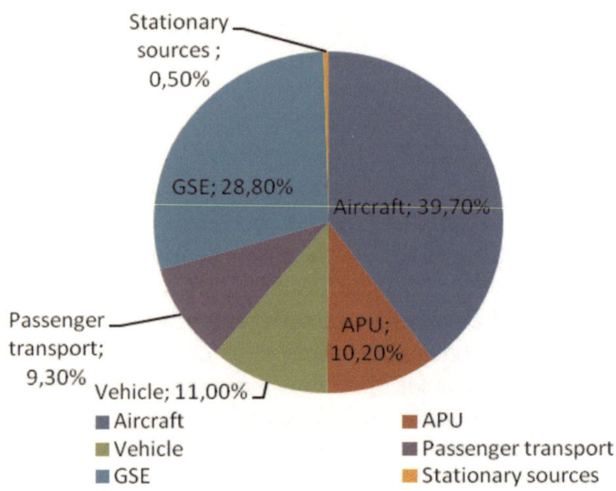

b

Figure 2.
The emissions inventory of NO_x [(a) annual emissions: 3.284 tons/year] and PM_{10} [(b) total emissions: 25 tons/year] within the Frankfurt International Airport for 2005 with an intensity of takeoffs and landings, 1300 per day.

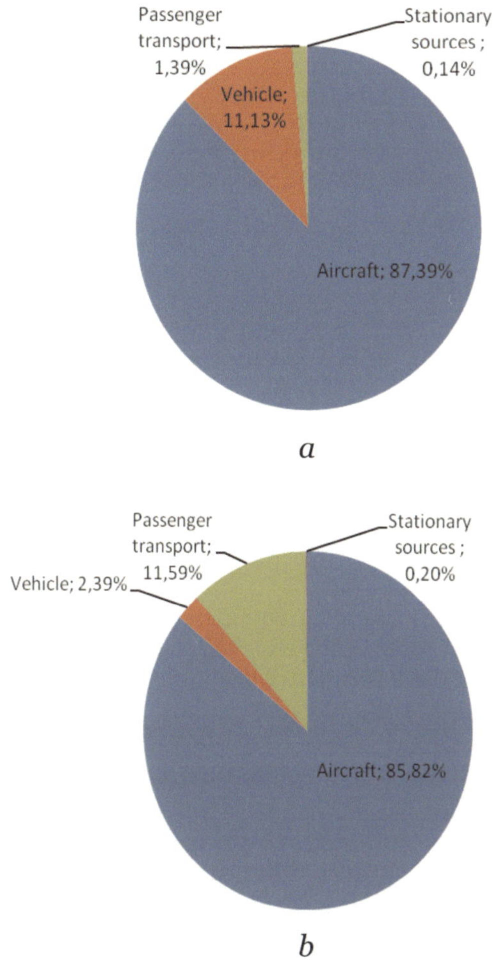

Figure 3.
The emissions inventory of NO_x (a) and PM10 (b) within Boryspil International Airport with an intensity oftakeoffs and landings 50,000 per year.

Ground-level emissions associated with the airport have the biggest impact on local air quality, whereas elevated aircraft emissions have less impact because they take place at increasing height. **Figure 1** shows aircraft produce approximately 54% of ground-level emissions, whereas airport-related traffic is estimated to emit a further 28% [5].

Analysis of inventory emission results at major European (Frankfurt am Main, Heathrow, Zurich, etc.) and Ukrainian airports highlighted that aircrafts (during approach, landing, taxi, takeoff and initial climb of the aircraft, engine run-ups, etc.) are the dominant source of air pollution in most cases under consideration [6, 9, 10], **Figures 2** and **3**. More than 50% of total NOx emissions inventory inside airport area is released by aircraft engines. As shown in **Figures 2(b)** and **3(b)**, the contribution of aircraft emission to total airport PM emissions is sufficiently high.

Considered problems are intensified in connection with rising tensions of expansion of airports and growing cities closer and closer each other (the most urgent is for Ukrainian airports, such as Zhulyany, Boryspil, Lviv, Odessa, and Zaporizhzhia) and accordingly growing public concern with air quality around the airport.

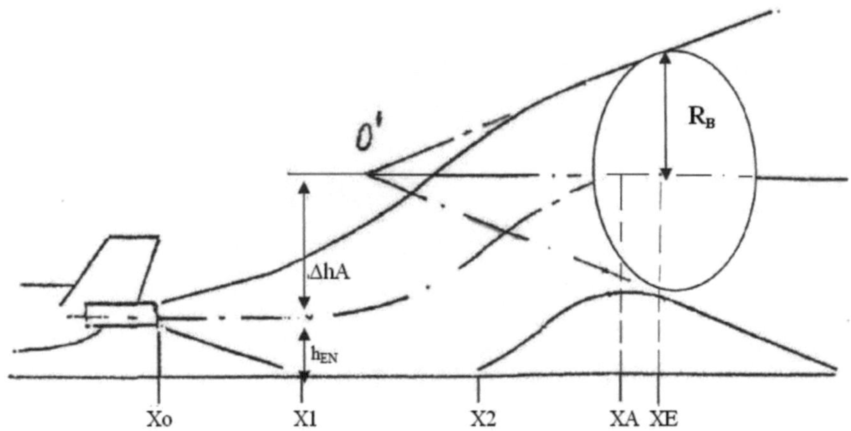

Figure 4.
Jet structure for jet transport model. Δh_A, X_A are the height and longitudinal coordinate of jet axis rise due to buoyancy effect, m; h_{EN} is the height of engine installation, m; R_B is the radius of jet expansion, m; X_1 is the longitudinal coordinate of first contact point of jet with ground, m; and X_2 is the longitudinal coordinate of a point of jet lift-off from the ground due to buoyancy effect, m.

Aircrafts are a special source of air pollution due to some features.

First of all, aircrafts are a moving (on the ground and in flight) pollution source with varying emission factors during landing and takeoff (LTO) as well as ground operation (engine start after maintenance and run-ups to check the correct operation of the flight system). At the airport, engine operation may change from idle to maximum thrust. Accordingly, temperature, exhaust gas velocity, and emissions of an aircraft engine may change within a wide range [11].

Second, the most important feature is the presence of a jet of exhaust gases, which can transport pollutants over rather large distances because of high exhaust velocities and temperatures (**Figure 4**). Such a distance is determined by the engine power setting and installation parameters, mode of airplane movement, and meteorological parameters. The results of jet model calculations show that depending on initial data, the jet plumes from aircraft engines range from 20 to 1000 m and sometimes even more [11].

So, to evaluate the aircraft contribution in Local Air Quality assessment of the airports accurately, it is important to take in mind few features of the aircraft during their landing-takeoff cycle (LTO), which define emission and dispersion parameters of the considered source.

2. Modeling of air pollution produced by aircraft engine emissions

Modeling of airport air pollution includes two parts: emission inventory and dispersion calculation.

ICAO Doc 9889 [12] recommends few tools for air quality analysis—to model emission inventory from every character groups of the spatially distributed sources as well as atmospheric concentrations resulting from emission dispersion: EDMS is based on Gaussian plume model (AERMOD) [13], LASPORT is based on Lagrangian particle model (LASAT) [14], and ALAQS-AV provides to use both Gaussian and Lagrangian approaches for dispersion calculations [12].

A complex model Pollution and Emission Calculation (PolEmiCa) for assessment of air pollution and emission inventory analysis, produced within the airport boundaries, has been developed at National Aviation University (Kyiv, Ukraine) [15]. It consists of the following basic components:

1. **engine emission model**—emission assessment for aircraft engines, including influence of operational factors;

2. **jet transport model**—transportation of the contaminants by the jet plume from the engine exhaust nozzle; and

3. **dispersion model**—dispersion of the contaminants in atmosphere due to turbulent diffusion and wind transfer.

2.1 Emission model

The emission inventory of aircraft emissions are usually calculated on the basis of certificated emission indexes, which are provided by the engine manufacturers and reported in the database of the International Civil Aviation Organization (ICAO) [16].

The emission indices rely on well-defined measurement procedure and conditions during aircraft engine certification. Under real circumstances, however, these conditions may vary and deviations from the certificated emission indices may occur due to the impact factors such as

- the life expectancy (age) of an aircraft—emission of an aircraft engine might vary significantly over the years (the average period is 30 years); usually aging aircraft/engine provides higher emission indices in comparison with same type but new ones;

- the type of an engine (or its specific modification, for example with different combustion chambers) installed on an aircraft, which can be different from an engine operated in an engine test bed (during certification); and

- meteorological conditions—temperature, humidity, and pressure of ambient air, which can be different for certification conditions.

So, the analysis of several measurement campaigns for idling aircraft at different European airports (London-Heathrow in 1999 and 2000, Frankfurt/Main in 2000, Vienna in 2001, and Zurich in 2003) [7] concludes that the largest difference between emission indices' measurement data and the ICAO data for CO for the RB211-524D4 engine was caused due to quite long life expectancy of B747-236 (aging aircraft and engines) (**Figure 5**). The oldest aircraft with an emission index of 52.9 g/kg was 25 years old; the other two were built in 1987 and 1983. Mean values

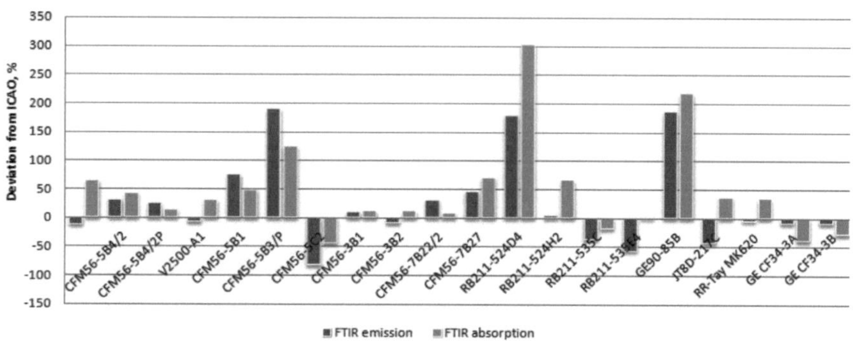

Figure 5.
Comparison measured EI_{CO} by FTIR emission and absorption spectrometry during measurement campaign for idling aircraft at the European airports.

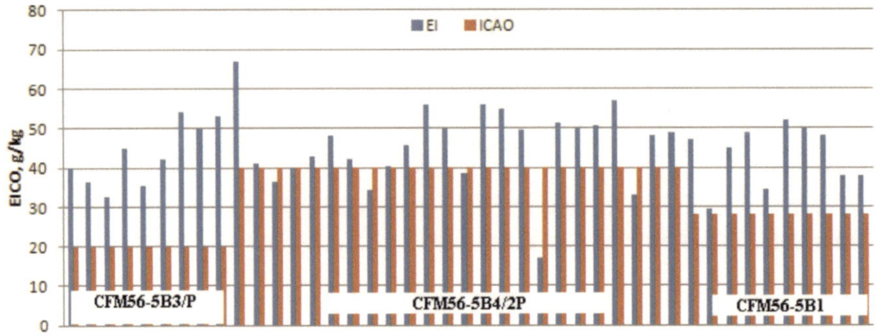

Figure 6.
Comparison EI_{CO} determined for CFM-5Bx engines with ICAO values for idling aircraft at European airports.

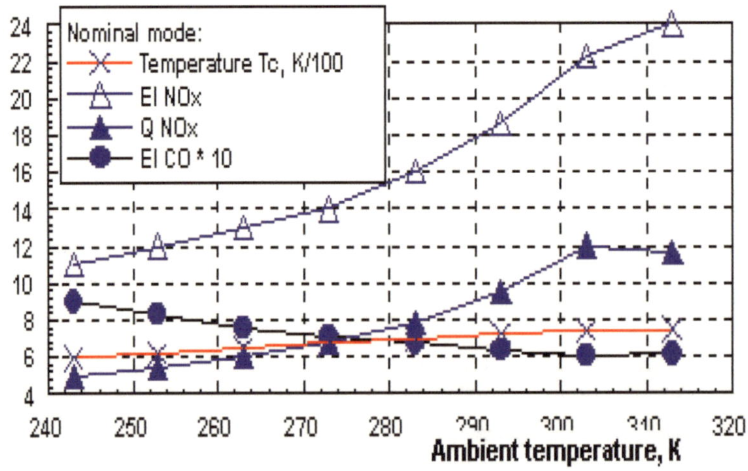

Figure 7.
Dependences of EE index EI [$g_{emission}/kg_{fuel}$], factor Q [g/s] for NO_x, and temperature behind the compressor T_c for D-36 engine from ambient temperature.

of the measured emission indices for three engine types (CFM56-5B1, CFM-5B4/2P, and CFM56-5B3/P) are nearly identical although the ICAO data of the CFM56-5B family differ by a factor of 2 (**Figure 6**) [7].

The dependences of engine thermodynamic parameters and EE index for NOx (assessed in g/kgfuel) and factor Q (assessed in g/sec) for the aircraft engine D-36 (installed on Yakovlev-42 and on Antonov-74, -148, and -158 aircrafts) are shown in **Figure** 7 as the functions from ambient temperature TA (basic engine control law for D-36 provides the constant value of compressor pressure ratio $\pi\sum^*$ in a broad range of ambient temperatures). Values of an emission index EI_{NOx} vary up to 50% in relation to value at International standard atmosphere conditions inside the range of ambient temperatures between −30 and + 30°C [17, 18].

A gradient of change of the factor Q_{NOx} at $T_A < T_{ALIM}$ is also large enough (with T_A growth, a factor Q_{NOx} monotonically increases). In case of change of engine automatic control mode at $T_A > T_{ALIM}$, the propellant consumption drops with the growth of temperature; therefore, monotonic character for Q_{NOx} dependence disappears and at $T_A > 30^{\circ}$C, the factor Q_{NOx} decreases [17, 18].

So, under operating conditions, engine emission characteristics are subject to changes as a result of influence of the meteorological factors.

Based on the obtained research outcomes of aircraft engine emission derivation, due to meteorological factor influences, the model was developed to recalculate the emission indices for ISA conditions EI_{iISA} into actual meteorological conditions EI_{it} [17, 18]:

$$K_{EIi} = \frac{EI_{it}}{EI_{iISA}} \tag{1}$$

For emission factor (in g/s or kg/hour), the recalculation into actual meteorological conditions are determined under the formula:

$$K_{Qi} = K_{EI_i} \cdot \left(\frac{T_t}{T_{ISA}}\right)^{1/2} \tag{2}$$

In **Table 1,** the correction coefficients for NOx emission factor KQnox and for products of incomplete fuel combustion KQco for average parameters of the engines while in operation are adduced.

Current calculation method, realized in software PolEmiCa, also implemented the recommendations of ICAO Doc9889 [12] for emission factor assessment, including the recommendations for aircraft engine emission.

The efficiency of the temperature (seasonal) factor account for pollution inventory produced by aircraft in airport area is shown in **Table 2** by matching the outcomes of calculation from previous and new calculation techniques [19].

2.2 Jet transport model

There are different types of engines installed on civilian aircraft currently: turbojet (TJE), turbofan (TFE), turboprop (TPE), and piston (PE). The process of contaminant transport by engine jet is described by the theory of turbulent jets [20]. The restrictions on the use of this theory are satisfied completely in the current task [21]: efflux from a jet engine is a very complex fast flow of hot gas, it is nonuniform, turbulent, and has various velocity scales and chemical reactions; the gas flow in jet is usually isobaric process, the pressure in the jet flow is equal to the atmospheric pressure, which is corresponding to the nature of incompressible flow; the Mach number of jet flow at outlet nozzle of the engine does not exceed 1; and the Reynolds number for the flow is rather large $U0D0/\nu > 10^5$, and the initial turbulence in the jet flow is quite moderate. For majority of the calculations, the simplifying preconditions were formulated and used: radial velocity profile has a self-preserving pattern; mechanisms of boundary layer formation near ground surface are not taken into account in this calculation; the external borders of a jet represent linear dependencies; the structure of shear layer is similar to free jet [11].

Temperature, °C	**−20**	**−10**	**0**	**+ 10**	**+ 20**
Factor K_{Qnox}	0,74	0,81	0,88	0,96	1,0
Factor K_{Qco}	1,3	1,2	1,1	1,04	1,0

Table 1.
Average values of aircraft engine emission factor recalculation into actual ambient temperature.

Techniques	**CO**	**HC**	**NO_x**	**PM**
Previous	307,000.1	104,200.	16,700.0	3400.0
ICAO LTO	282,754.6	97,139.2	18,621.1	2859.4
Actual LTO for considered airport	185,055.1	59,556.4	16,869.1	2207.3
Actual LTO for considered airport + temperature factor	190,246.1	61,254.1	15,984.1	2207.3

Table 2.
Calculated aircraft engine pollution, kg.

The conditions of jet outflow define the type of its physical model and appropriate algorithm of its parameters calculation. The choice of the model depends on the direction of the jet at exhaust nozzle relative to the direction of the wind and/or airplane motion and from the speeds of the jet, airplane, and wind. The initial parameters for jet calculations are: slipstream flow parameter m = U_H/U_0, where U_0 is the velocity of the jet at engine nozzle, m·s^{-1} and U_H is the speed of an external air flow, m·s^{-1}; $U_H = U_W + U_{PL}$, where U_W is the wind speed, m·s^{-1} and U_{PL} is the airplane speed, m·s^{-1}; N_{en} – number of the engines in operation, angle between vectors of wind and jet speeds ψ, grad. For ground stages of LTO cycle in airport area, the slipstream parameter m < < 1; therefore, in most cases, it is possible to take advantage of semiempirical modeling of the nonisothermal-free jets.

Turbulent-free jet can be divided into three stages: initial (potential core), transitive (flow development region), and developed (fully developed flow) [20]. Their boundaries along the length of jet axis *S* and their expansion *R* (on considered sites) are defined by the formulas [11, 20, 22]:

- **for an initial stage**:

$$\overline{S}_{IN} = (11.5 - 3.5 \times Q_T) \times (1 + 2.5 \times m); \overline{R}_{IN} = 0.27 \times \overline{S}_{IN} \qquad (3)$$

- **for a transitive stage**:

$$\overline{S}_T = 1.5 \times \overline{S}_{IN}; \overline{R}_T = 1.5 \times \overline{R}_{IN} \qquad (4)$$

- **for the fully developed stage**:

$$\overline{S}_B = 12.4 \times Q_T^{-1/2} \times (1 - m)/m + \overline{S}_T; \overline{R}_B = 2.728 \times (Q_T m)^{-1/2} + \overline{R}_T \qquad (5)$$

where $\overline{S} = {}^{S}/_{R_0}$ and $\overline{R} = {}^{R}/_{R_0}$; R_O is the radius of engine exhaust nozzle; *m* is the slipstream flow parameter; and $Q_T = T_0/T_A$, where T_0 and T_A are the temperature of the jet and atmosphere, K. Parameter Q_T for modern engines changes within the limits of 1.15–2 for the operational settings of engine power.

The stage of a jet, which is defined by boundary S_B (6), determines a point (X_E, Y_E, Z_E) on a jet axis, where centerline flow speed U_m and the wind speed U_W become equal. From this point, it is assumed that atmospheric turbulence and wind play a dominant role in the plume behavior and its further dispersion, while the jet parameters influence is not already sufficient at this stage of plume development.

At point (X_E, Y_E, Z_E), a jet center-line due to buoyancy effect takes height of plume rise (it is equal to effective height of source *H* in (1) and (2)) [11, 20, 22]:

$$Z_E = h_{EN} + \Delta h_A, \qquad (6)$$

where h_{EN} is a height of engine installation (of their axis above a ground surface), m and Δh_A is a height of jet rise, m.

For an estimation of the buoyancy characteristics, the Archimedes number is introduced:

$$Ar_0 = {}^{2 \times g \times R_0 \times (Q_T - 1)}/_{U_0^2}, \qquad (7)$$

The height of the jet is given by the empirical relationship [23]:

$$\Delta h_A = 0.013 \times Ar_0 \times \overline{X}_A^3 \times R_0, \qquad (8)$$

where $\overline{X_A} = {}^{X_A}/_{R_0}$, X_A is the longitudinal coordinate of jet axis curved by buoyancy effect, m (**Figure 1**) and can be calculated by the following formula:

$$\overline{X}_A = \left\{ \left(1 + 0.156 \times Ar_0 \times \overline{S}_A^2\right)^{1/2} - 1/0.078 \times Ar_0 \right\}^{1/2}. \tag{9}$$

The concentration is changed along the length of jet in dependence with its type. Taking into account that flow parameter *m* in jet is rather small, the concentration *C* of the contaminant on a surface (*X, Z*) is defined as [20, 22]:

$$C = \frac{2 \times C_0 \times K_C \times K_E}{Q_T^{1/2} \times X} \times \left[1 - \left(\frac{Z}{0.23668 \times X}\right)^{1.5}\right]^{1.5}, \tag{10}$$

where C0 is the concentration at the exhaust nozzle of the engine, $\mu g \cdot m^{-3}$; K_C = 9.5 for the free jet, K_C = 6.5—for an opposite jet; and K_E takes into account influence of a reflecting surface on straightline characteristics of a jet: $\hat{h}_{EN} < 20$ K_E = 1–0.025hEN, at $\hat{h}_{EN} \geq 20$, K_E = 1, where $\hat{h}$ = h/R0.

Considered version of complex model PolEmiCa is based on a semiempirical model of turbulent jets and not taking into account ground surface impact on jet structure and its behavior [11]. It was argued that development of three-dimensional model of exhaust gases jet from aircraft engine near the ground is an important research topic for airport LAQ [24–26].

A three-dimensional model of a jet was generated in Fluent 6.3 by using large Eddy simulation (LES) method to reveal the unsteady ground vortices and turbulence characteristics of fluid flow, to investigate transient parameters of hot gases in jet and their dispersion.

The jet from aircraft engine exhaust near ground surface is corresponding to a wall jet if an aircraft is moving on this surface. Numerical simulation of wall jets was performed in Fluent 6.3 for engine NK-8-2 U of the aircraft Tupolev-154 for different operational conditions.

For the considered task, a computational domain was built to simplify the problem and optimize the mesh distribution where it is needed mostly (i.e., near the engine exhaust and ground surface) (**Figure 8**).

The zone of ground vortices formation—between ground surface and aircraft engine exhaust nozzle—is characterized by structured mesh with higher resolution, with an aim to investigate the ground vortices generation processes and basic mechanisms of boundary layer formation, ground surface impact on fluid flow mechanics, and particularly Coanda effect occurrence. Zone of engine nozzle exhaust is discretized using a very fine structured mesh to capture the jet development pattern and its vortices structure [24, 25].

For considered task, the boundary conditions were specified to the boundaries of the computational domain of jet flow field (**Figure 9**).

LES provides an approach inside which large eddies are explicitly resolved in time-dependent simulation using low-pass-filtered Navier-Stokes equations [25]. Smagorinsky's subgrid model was set to model the smaller eddies (fluctuation component of instantaneous velocity of modeling fluid flow) that are not resolved in the LES. All the calculations were made with a second-order discretization.

Comparison of results from numerical simulations of free and wall jets for engine idle operation (U0 = 50 $m \cdot s^{-1}$; T0 = 343 K) revealed some differences in their structures and properties.

Axial velocity profiles based on Fluent 6.3 results show (**Figure 10**) a substantial difference between the wall and free jet. First, the decay rate is 40–50% higher for

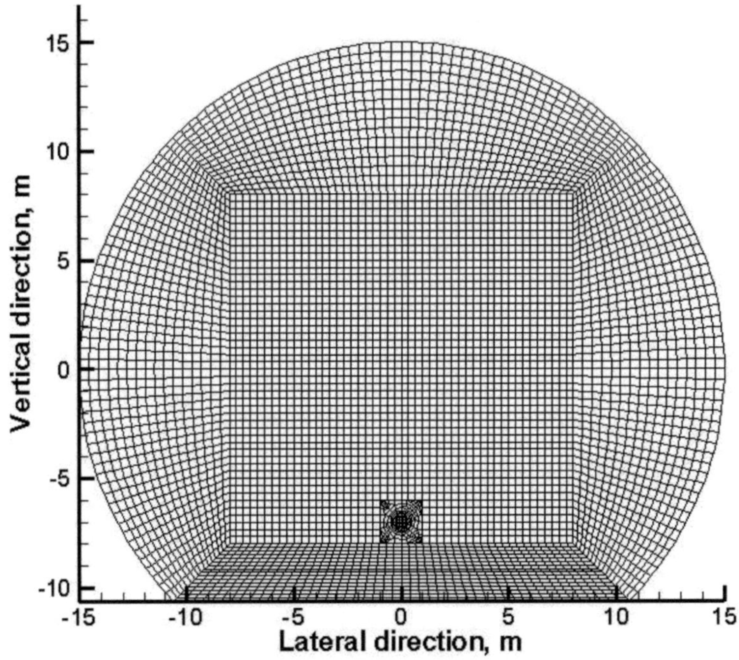

Figure 8.
Geometry model and computational mesh visualization in vertical plane.

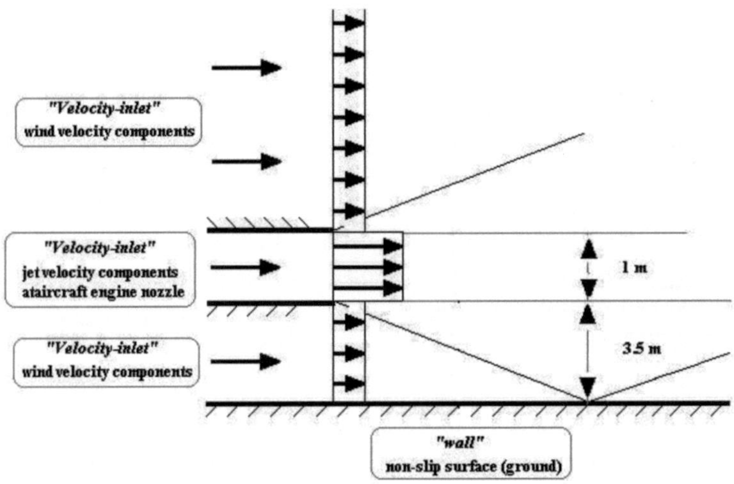

Figure 9.
Boundary conditions for CFD simulations of exhaust gases of jet from aircraft engine near ground.

free jet than for the wall jet. In the case of wall jet, the maximum velocity is high and equal to 50% of initial velocity at a distance of 90 diameters of the jet penetration, whereas the free jet is relatively slow and equal only to 10% of the velocity at exhaust nozzle of the engine, **Figure 10**. Second, the wall jet penetrates deeper ($S_{Bwall} \approx 150$ m) than the free jet ($S_{Bfree} \approx 100$ m) (**Figure 11**). As shown in **Figure 12,** jet arises over the ground surface due to buoyancy effect much faster (longitudinal coordinate, X_A = 65 m) and higher for free jet (height of plume rise, ΔhA = 17.8 m), than in case of wall jet (X_A = 135 m, Δh_A = 14 m).

The same differences in the structure and properties of free and wall jets were revealed for different operational conditions (U0 = 100 m·s^{-1}; T0 = 343 ÷ 673 K).

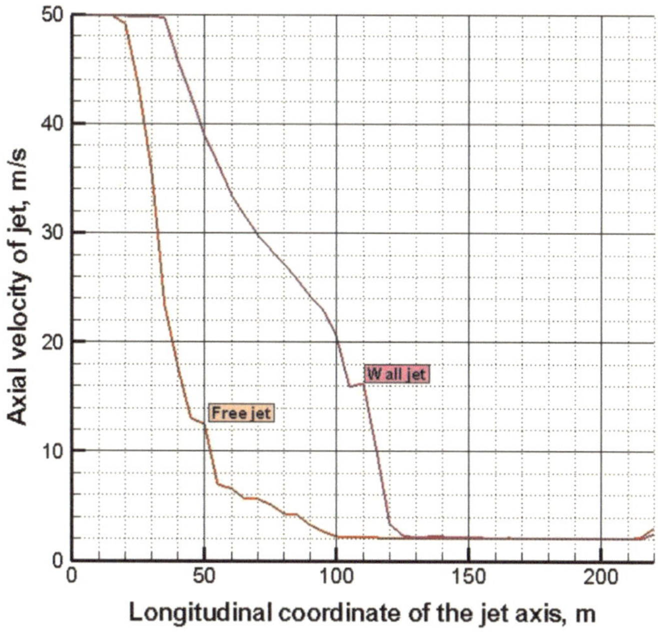

Figure 10.
Maximum velocity decay along the axis of the free and wall jets.

The ground surface sufficiently impacts on jet's structure and behavior. Numerical simulations of wall jet by Fluent 6.3 defined a decrease of buoyancy effect of height rise, which is 3–5 times less (**Figure 13a**) and an increase of longitudinal coordinate of jet penetration by 30%, (**Figure 13b**).

Comparison of the calculated parameters of the jet (height and longitudinal coordinate of jet axis arise due to buoyancy effect, length of the jet penetration) by Fluent 6.3 and semiempirical model for aircraft engine jets implemented in complex model PolEmiCa proves the found trend of the jet behavior. Thus, the including the ground impact on the jet structure and its behavior by Fluent 6.3, provides longitudinal coordinate increase and height reduction of buoyancy effect.

2.3 Dispersion model

The basic model equation for definition of instantaneous concentration C at any moment t in point (x,y,z) from a moving source from a single exhaust event with preliminary transport by jet on distance X_A and rise on total altitude H (**Figure 4**) and dilution of contaminants by jet (σ_0) has a form [11, 19]:

$$C(x,y,z,t) = \frac{Q\, exp\left[-\dfrac{(x\text{-}x')^2}{2\sigma_{x0}^2 + 4K_x t} - \dfrac{(y\text{-}y')^2}{2\sigma_{y0}^2 + 4K_y t}\right]}{\left\{8\,\pi^3\left[\sigma_{x0}^2 + 2K_x t\right]\left[\sigma_{y0}^2 + 2K_y t\right]\right\}^{1/2}} \times \left\{\frac{exp\left[-\dfrac{(z\text{-}z'\text{-}H)^2}{2\sigma_{z0}^2 + 4K_z t}\right] + exp\left[-\dfrac{(z + z' + H)^2}{2\sigma_{z0}^2 + 4K_z t}\right]}{\left[\sigma_{z0}^2 + 2K_z t\right]^{1/2}}\right\} \tag{11}$$

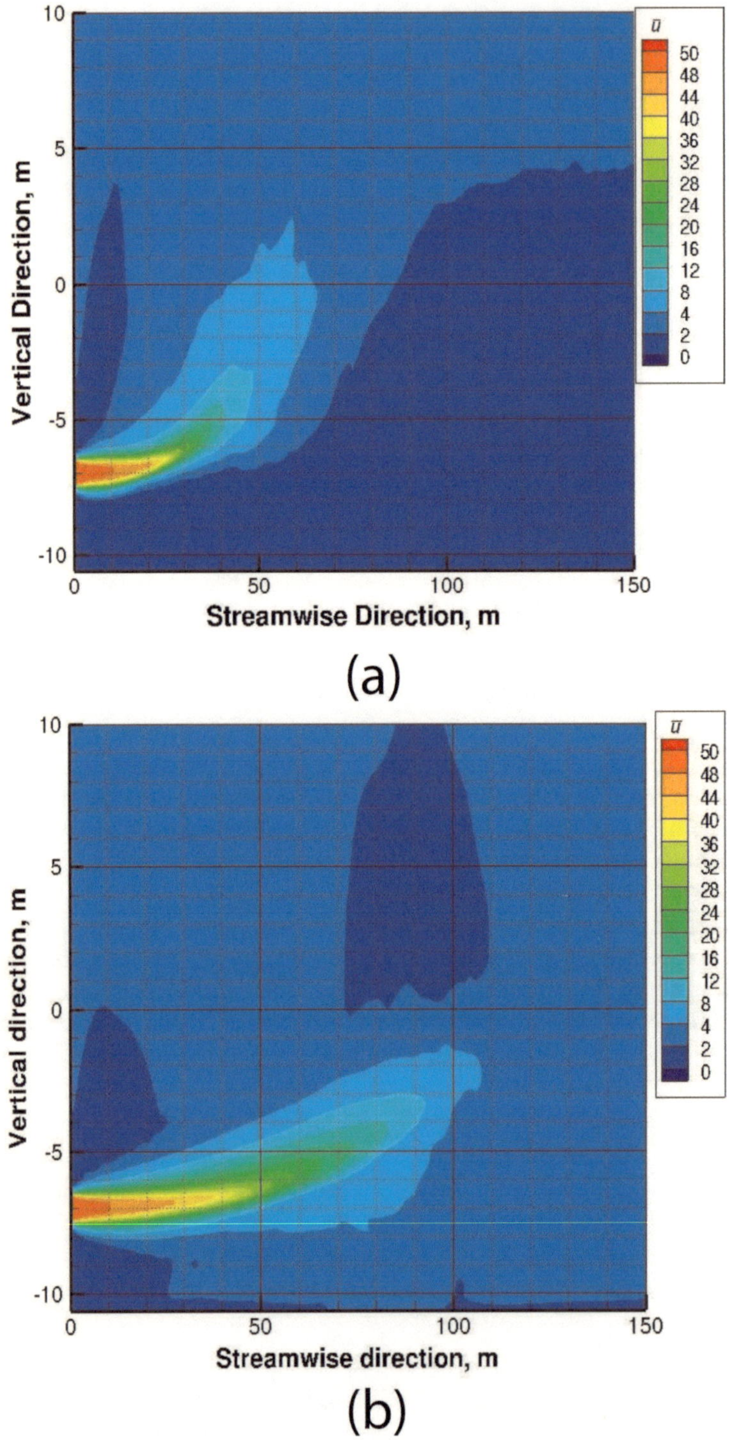

Figure 11.
Mean velocity contours for (a) free jet and (b) wall jet in streamwise direction after 10 s.

where K_X, K_Y, K_Z are the diffusion factors ($m^2 \cdot s^{-1}$) for atmosphere turbulence along three axes, axis *OX* is directed along wind direction. Aircraft is considered as a moving emission source, thus current coordinates (x', y', z') of the emission source in movement during time t' are defined as:

$$
\begin{aligned}
x' &= x_0 + u_{PL}t' + 0.5a_{PL}t'^2 + u_w(t + t'); \\
y' &= y_0 + v_{PL}t' + 0.5b_{PL}t'^2; \\
z' &= z_0 + w_{PL}t' + 0.5c_{PL}t'^2.
\end{aligned}
\tag{12}
$$

where x_0, y_0, z_0 are initial coordinates of the source, m; u_{PL}, v_{PL}, w_{PL} are vector components of source speed, m·s^{-1}; a_{PL}, b_{PL}, c_{PL} are vector components of source acceleration, m·s^{-2}; and u_w is the wind speed, m·s^{-1}.

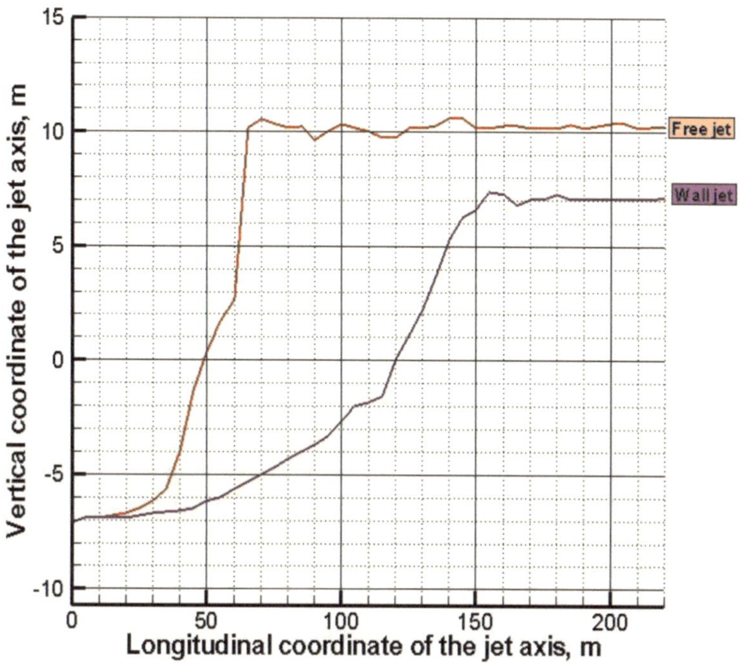

Figure 12.
Buoyancy effect of free and wall jets: longitudinal and vertical coordinates of jet axis.

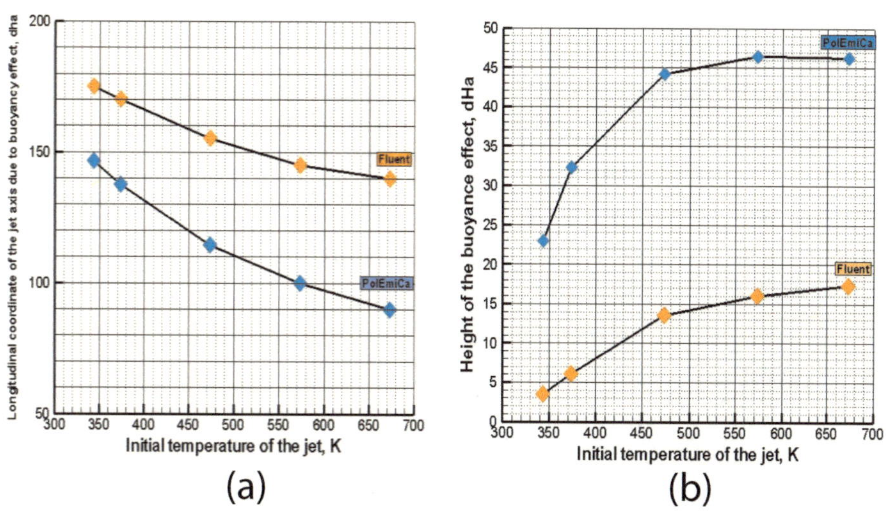

Figure 13.
Comparison of buoyancy effect parameters calculated by Fluent 6.3 and complex model PolEmiCa: longitudinal coordinate (a) and height of jet rise (b).

According to considered formula (11), a dispersion model integrates engine emission model and jet transport model via including the following parameters:

Q—emission rate is provided by engine emission model and includes influence operational and meteorological conditions [17–19].

H—height of buoyancy effect and horizontal σ^2_x, σ^2_y and vertical σ^2_z dispersion are provided by *jet transport model* [24–26].

In other words, engine emission model and jet transport model provide input data to calculate concentration values by the dispersion model.

The development of three-dimensional model of wall jet by using CFD tool (Fluent 6.3) allows to include the ground impact on basic parameters of the exhaust gases jet (i.e., plume buoyancy effect, length, and dispersion characteristics) for further dispersion modeling (11). It may be concluded that using the CFD tool allows us to improve the PolEmiCa model by taking into account the impact of ground surface on the jet structure and its behavior. So, it means that the improvement is achieved with input parameters for further dispersion calculation.

3. Measurement of air pollution produced by aircraft engine emissions

The verification of the PolEmiCa model with measurement data was done initiatively for trials made in airports of Athens (Greece, 2007) [27] and Boryspil (Ukraine, 2012) [28]. In both cases, the comparisons were quite good, showing appropriate correspondence of the model to subject of assessment.

Comparison between calculated and measured NOx concentrations (averaged for 1 min) in aircraft engine plume under real operation conditions (aircraft accelerating on the runway during takeoff stage of flight) at Athens airport is shown in **Table 3** and **Figure 14**.

Besides, results were defined for the cases with and without jets from the engines to show that with jets, they are more equal (by 17%) to measured data,

№	Aircraft	Engine	Calculated concentration		Measured concentration	
			NOx (delta), μg/m³		NOx (delta), μg/m³	
			With jet	Without jet	Value	Error
1	B737-3YO	CFM56-3C1	27,43	30,01	31,8	3,2
2	B737-3Q8	CFM56-3B2	30,7	33,50	28,0	2,8
3	B737-45S	CFM56-3B2	29,76	27,95	23,6	2,4
4	B737-4Q8	CFM56-3B2	31,28	34,93	56,9	5,7
5	A-310	CF6-80C2A8	88,86	122,12	86,1	8,6
6	A-319	CFM56-5B5	29,85	32,27	26,9	2,7
7	B747–230	CF6-50E2	163,63	205,37	82,5	8,2
8	A-321-211	CFM56-5B-3	81,78	89,74	43,3	4,3
9	A320–214	CFM56-5B-4	49,99	52,29	16,4	1,6
10	B737-33A	CFM56-3B1	25,5	27,95	11,5	1,1

Table 3.
Measurement results by TE42C-TL96 system and calculation results by PolEmiCa model of NOx concentration in plume from aircraft engine emission for maximum operation mode.

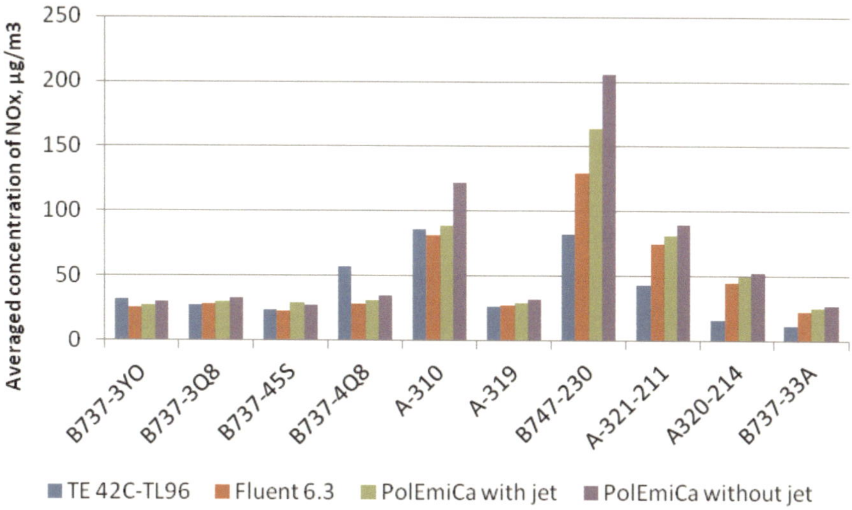

Figure 14.
Comparison of measured and modeled averaged concentrations of NOx (for a period of 1 min) under takeoff conditions (maximum operation mode of aircraft engine).

because impact of jet basic parameters (buoyancy effect and dispersion characteristics) on concentration distribution was estimated by complex model PolEmiCa (**Table 3** and **Figure 14**). Comparison between measurements and the PolEmiCa/ Fluent 6.3 model is significantly better (by 20%), because lateral wind and ground impact on jet parameters (height of buoyancy effect, jet length penetration, and plume dispersions) were included in the model.

The better agreement was obtained between the calculated and measured instantaneous concentration (averaged for 3 s) in aircraft engine jet under real operation conditions (aircraft accelerating on the runway and takeoff) at Boryspil airport.

As shown from **Table 4** and **Figure 15**, the modeling results for each engine are in good agreement with the results of measurements by the AC3 2 M system due to taking into account the jet- and plume regime during experimental investigation at Boryspil airport. Also, using CFD code (Fluent 6.3) allows to improve results by 30% (coefficient of correlation, r = 0.76) by taking into account lateral wind and ground impact on jet parameters.

Aircraft	Aircraft engine	ELAN		AC3 2 M			PolEmiCa CFD (Fluent 6.3)		PolEmiCa	
		Peak 1	Peak 1	Background	3 м	6 м	1 engine	All engines	1 engine	All engines
		NOx	NOx	NOx	NOx	NOx	NOx	NOx	NOx	NOx
BAE147	LY LF507-1H	38	35	1,70	22,067	33,9	35,1	70,46	48,9	202,3
A321	CFM56-5B3/P	39	39	0,72	44,00	54,2	90,85	182,90	184,2	371,2
B735	CFM-563C1	40	45	0,77	94,095	76,57	60,03	120,91	35,3	71,10
B735	CFM56-3B1	45	41	1,74	29,20	23,4	42,34	85,30	33,7	67,76

Table 4.
Comparison measured (AC3 2 M, ELAN) and calculated concentration (averaged for 3 s) of NOx produced by aircraft engine emissions at accelerating stage on the runway.

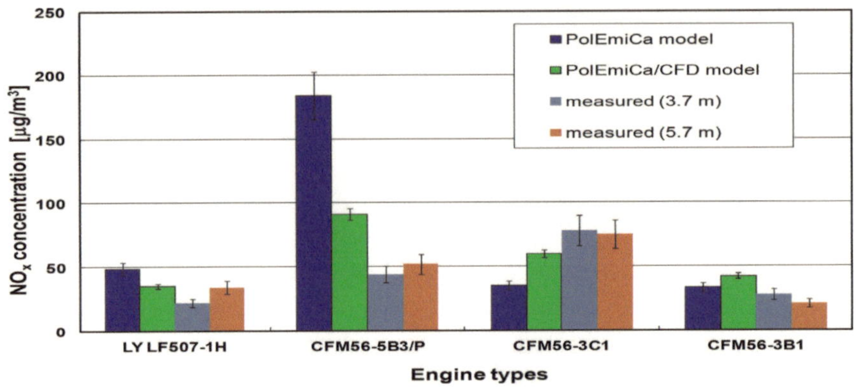

Figure 15.
Comparison of the PolEmiCa and PolEmiCa/CFD model results with the measured NOx concentration at different heights for selected aircraft engines under maximum operation mode.

4. Conclusions

Analysis of inventory emission results at the major European and Ukrainian airports highlighted that aircrafts (during approach, landing, taxi, takeoff and initial climb of the aircraft, engine run-ups, etc.) are the dominant source of air pollution in most cases under consideration. The aircraft is a special source of air pollution. Thus, the method for LAQ assessment of the airports has to take in mind few features of the aircraft during their landing-takeoff cycle (LTO), which defines emission and dispersion parameters of the considered source.

CFD numerical simulations of aircraft engine exhaust jet near to ground surface show that structures, properties, and fluid mechanics of jets are influenced by the ground surfaces, providing longer penetration, less rise, and appropriate dispersion parameters of the jets, and accordingly little bit higher concentrations of air pollution. So, using results obtained from CFD simulations (Fluent 6.3) of aircraft engine jet dynamics allow us to improve LAQ modeling systems (improved version of PolEmiCa).

Comparison of measured and modeled NOx concentrations in the plumes from aircraft engines was significantly improved (by 20%—at Athens and by 30%—at Boryspil airports) by taking into account lateral wind and ground impact on jet parameters (height of buoyancy effect, jet length penetration, and plume dispersions).

Author details

Oleksandr Zaporozhets and Kateryna Synylo*
National Aviation University, Kyiv, Ukraine

*Address all correspondence to: synyka@gmail.com

References

[1] EEA. 2016. European Aviation Environmental Report. Available from: https://ec.europa.eu/transport/sites/transport/files/european-aviation-environmental-report-2016-72dpi.pdf [Accessed: 29 January 2016]. ISBN: 978-92-9210-197-8

[2] Carslaw D, Beevers S, Ropkins K, Bell M. Detecting and quantifying aircraft and other on-airport contributions to ambient nitrogen oxides in the vicinity of a large international airport. Atmospheric Environment. 2006; **40**(28):5424-5434

[3] Peace H, Maughan J, Owen B, Raper D. Identifying the contribution of different airport related sources to local urban air quality. Environmental Modelling and Software. 2006;**21**(4): 532-538

[4] Jung K, Artigas F, Shin J. Personal, indoor, and outdoor exposure to VOCs in the immediate vicinity of a local airport. Environmental Monitoring and Assessment. 2011;**173**(1):555-567

[5] Nicholas Rand. Heathrow Air Quality Strategy, 2011–2020. Available from: https://www.heathrow.com/file_source/Company/Static/PDF/Communityandenvironment/air-quality-strategy_LHR.pdf

[6] Celikel A, Duchene N, Fleuti E, Fuller I, Hofman P, Moore T, and Silue M. Airport local air quality studies: Zurich Airport Emissions Inventory Using Three Methodologies. 2004. Available from: https://www.eurocontrol.int/eec/public/standard_page/DOC_Conf_2005_007.html [cited: July 2005]

[7] Schäfer K, Jahn C, Sturm P, Lecher B, Bacher M. Aircraft emission measurements by remote sensing methodologies at airports. Atmospheric Environment. Elsevier; 2003;**37**:5261-5271

[8] Herndon S, Jayne J, Lobo P. Commercial aircraft engine emissions characterization of in-use aircraft at Hartsfielde Jackson Atlanta international airport. Environmental Science & Technology. 2008;**42**(6): 1877-1883

[9] Fraport Environmental Statement (2014). Including the Environmental Program, until 2017. Fraport AG; 2015. pp. 24-30

[10] Umweltbericht. Umwelterklärung 2008 mit Umweltprogramm bis 2011 für den Standort Flughafen Frankfurt Main. Fraport AG; 2008. pp. 100-104

[11] Zaporozhets O, Synylo K. PolEmiCa—tool for air pollution and aircraft engine emission assessment in airport. In: The Proceedings of the Second World Congress "Aviation in the XXI-st Century"; Ukraine: Kyiv; 2005. pp. 4.22-4.29

[12] ICAO Doc9889. AirportAirQuality. 1st ed. INTERNATIONAL CIVIL AVIATION ORGANIZATION 999 University Street, Montréal, Quebec, Canada; 2011. 200 p

[13] Emissions and Dispersion Modelling System (EDMS). Reference Manual. FAA-AEE-01-01. U.S. Department of Transportation Federal Aviation Administration. Washington, D.C.: CSSI, Inc; 2002

[14] Janicke Consulting. LASPORT version 1.3 Reference Book; 2005. 93 p

[15] Zaporozhets O. 2015. PolEmiCa Local Air Quality Model Evaluation/O. In: Zaporozhets K. Synylo. ABIA2015: XII міжнародна наук.-техн. конф., 2015: збірник матеріалів—К., – С.29.12–29. 15. (19)

[16] ICAO data Bank of Aircraft Engine Emissions. Montreal: ICAO. Doc. 9646-AN/943; 1995. 152 p

[17] Zaporozhets O, Synylo K. Improvements on aircraft engine emission and emission inventory assessment inside the airport area. Energy Journal (SCOPUS). 2017;**5**: 1350-1357

[18] Zaporozhets O, Synylo K. Operational conditions influence on aircraft engine emission and pollution inside the airport. International Journal of Sustainable Aviation. 2017;**3**(1):1-17

[19] Synylo K, Zaporozhets O. PolEmiCa model for local air quality assessment in airports. Community Modeling and Analysis System, 16th Annual CMAS Conference; 23–25 October, 2017. Chapel Hill, NC, USA: University of Northern Carolina. 6 p

[20] Abramovich GI. The Theory of Turbulent Jets. Moscow: Physmatgiz; 1960. 716 p

[21] Tokarev VI, Zaporozhets OI. Mathematical principles of modeling of atmospheric contamination processes by effluents of harmful substances at maintenance of aircrafts. In: State and Perspective of Activities on Environment Protection in Civil Aviation. Moscow: GosNIIGA; 1987. pp. 53-60

[22] Zaporozhets O. Evaluation of preliminary dispersion of effluents of contaminant substances by jets. In: Problems of Environment Protection at Intesification of Aircraft Productions. Kyiv: KIECA; 1986. pp. 42-51

[23] Taliev VN. Aerodynamics Ventilation—M; 1979. 295 p

[24] Pope Turbulent Flows. Cornell University, New York: Cambridge University Press; 2000. 841 p

[25] Zaporozhets O, Synylo K, Fröhlich J, Stiller J. Improvement of airport local air quality modeling. Journal of Aircraft. 2017;**54**(5):1750-1759

[26] Zaporozhets O, Synylo K. New and improved LAQ models for assessment of aircraft engine emissions and air pollution in and around airports. On board a sustainable future: ICAO Environmental Report 2016, International Civil Aviation Organization 999 University Montreal, QC, Canada H3C 5H7. 2016. 250p. pp. 82-84

[27] Zaporozhets O, Wiesen P, Kurtenbach R, Synylo K. Comparison between modelled and measured NOx concentrations in aircraft plumes at Athens international airport. International Journal of Sustainable Aviation. 2017;**4**(3):1-17

[28] Zaporozhets O, Wiesen P, Kurtenbach R, Synylo K. Measurement of aircraft engine emissions inside the airport. In: The 4th International Conference on Transport, Atmosphere and Climate; June 2015. Bad Kohlgrub (Germany): Proceedings; 2015. pp. 28-33

Chapter 6

Sonic Boom Mitigation through Shock Wave Dispersion

Constantin Sandu, Radu-Constantin Sandu and Cristian-Teodor Olariu

Abstract

Lately, the interest for passenger space planes, supersonic passenger aircraft, and supersonic business jets has greatly increased. In order to mitigate the sonic boom effects at ground level, some aerospace companies proposed airplanes that have a very small transversal fuselage section or that have a curved ("shaped") fuselage. Obviously, shaping the fuselage leads to the increase of dynamic drag and manufacturing cost. Reducing the fuselage transverse section leads to reducing the useful volume inside fuselage and increases the landing distance of aircraft. The solution presented in this chapter shows that it is theoretically and technologically possible as the shock wave to be dispersed through mechanical or electrical means. The shock wave is in fact a stationary effect generated by the move of aircraft with constant speed relatively to surrounding air. If this feature is in a way or another canceled, the shock wave is dispersing. Due to dispersion of the shock wave the 'N' wave at the ground is tens of times larger and the sonic boom is correspondingly lower. The shock wave dispersion system of the future could be mechanical or electrical is activated only when the supersonic aircraft/space plane is flying horizontally over community.

Keywords: sonic boom mitigation, shock wave dispersion, supersonic aircraft, supersonic business jet, space plane

1. Introduction

The first manned airplane, which exceeded the speed of sound in horizontal flight was the American airplane X-1 manufactured by Bell Aircraft Corporation [1]. On the 14th of October 1947, the X-1 aircraft was air-launched at the altitude of 7000 m from the bomb bay of a Boeing B-29 and then climbed to the test altitude of 13,000 m. Piloted by Chuck Yeager, the aircraft reached a speed of 1127 km/h (Mach = 1.06) in horizontal flight. Since the maiden flight, the aircraft accumulated a number of 78 flights and on the 26th of March, 1948 it attained a speed of 1540 km/h (Mach = 1.45) at the altitude of 21,900 m.

Because at that time no jet engine was powerful enough, the aircraft Bell X-1 was powered by a four-chamber XLR-11 rocket engine that produced a static thrust of 26.5 kN. This was the first time when the sonic boom was revealed as a natural phenomenon generated by the aircraft breaking the sound barrier. In essence, the sonic boom is the manifestation of the shock waves generated by a supersonic aircraft perceived at ground level.

After this important event, a multitude of supersonic aircraft having exclusive military applications was developed and manufactured in series by the most technologically advanced countries. Simultaneously, the phenomenon of sonic boom was intensively researched from a theoretical and experimental point of view [2–6].

At the beginning, due to the fact that the interest for the supersonic flights was exclusively for military applications, the ecological impact of sonic boom was not taken into account.

However, on the 2nd of March 1969, the first flight of Concorde supersonic passenger aircraft took place. This aircraft was produced by the French company Aerospatiale and British Aircraft Corporation (BAC). Concorde was a large enough aircraft (much larger than a usual military supersonic aircraft) to reveal the extremely high negative environmental impact of the sonic boom [7]:

Length: 62.19 m
Wingspan: 25.6 m
Height: 12.19 m
Empty weight: 79,260 kg
Capacity: max. 144 passengers
Maximum speed: Mach 2.04 (≈2179 km/h) at cruise altitude
Cruise speed: Mach 2.02 (≈2158 km/h) at cruise altitude
Range: 7222.8 km
Service ceiling: 18,290 m

On the 24th of October 2003, Concorde operated its last flight, leaving the aircraft market and airspace. An important reason was the impact of the sonic boom produced on the environment/community.

This fact raised the interest for sonic boom mitigation. Thus, important papers [8–11] were written on this subject, and a number of solutions for sonic boom mitigation were filled in patent [12–18].

Lately, an important change on the aircraft market took place: the start of a high demand for supersonic business jet and a continuous rise of interest for very high-speed passenger transportation, supersonic and hypersonic airliners.

An important problem generated by supersonic aircraft is the effect of sonic boom at the ground level. The sonic boom is an "N"-shaped pressure distribution, which spans the ground when an aircraft is flying at supersonic speed. The lower the flying height, the higher the material damages and annoyance produced in community.

This problem blocked the development of supersonic civil aircraft for a long period of time.

The state of the art regarding the solutions for mitigation of sonic boom effects at ground level is presented in Chapter 2 together with the drawbacks of these solutions.

In the next points of this chapter, the authors underline some important characteristics of shock wave, which support a new possible solution to mitigate the sonic boom impact at ground level: dispersion of shock wave mainly through vibration of aircraft nose surface and wing leading to edge surface. The explanation is simple: the shock wave is a steady-state effect, which is generated through moving of aircraft with a constant speed. If this steady-state characteristic of flight is canceled through vibration of the specified surfaces, the shock wave is dispersed, and its effect at ground level (known as "sonic boom") is greatly reduced.

2. The state of the art

For reducing sonic boom effects at ground level, companies as Supersonic Aerospace International, Lockheed, in collaboration with NASA, Boeing, Airbus,

Dassault, and Aerion Corporation proposed airplanes having thin or curved (shaped) fuselages, and other designers proposed biplane type aircraft.

Some design solutions are presented in **Figures 1** and **2** [19]. The long aircraft having a small cross section (**Figure 1**) needs a too long landing distance, and the space for passengers inside fuselage is small. Nevertheless, it seems that this solution began to be preferred at present by aircraft manufacturers. This preference is explained by the manufacturing costs that are low because no major change in the current technology is necessary.

Obviously, the curved (shaped) fuselage (**Figure 2**) strongly perturbs the airstream flowing around the aircraft. As a result, more power is required for flight. At the same time, the curved fuselage considerably increases the manufacturing costs of aircraft.

Figure 1.
An advanced Lockheed Martin concept [19].

Figure 2.
An advanced Northrop Grumman concept [19].

For a very long period of time, the "shaping" solution was the preferred one. According to this solution, shaping the fuselage leads to the changing of the "N" wave shape at ground level and mitigation of its impact.

The theory of sonic boom mitigation through shaping was established during the 1960s–1970s with the papers written by Seebass, Carlson, and Darden [8, 20, 21]. This theory was not proven until 2002.

In 2002, the Defense Advanced Research Projects Agency (DARPA) selected several companies for the Phase II of the Quiet Supersonic Platform (QSP) program [22]. The allocated research funds were of about 9 million USD. The selected companies were the following:

- Lockheed Martin, Advanced Development Company, Palmdale, California
- Northrop Grumman Corporation, El Segundo, California
- Arizona State University, Tempe, Arizona
- General Electric, Cincinnati, Ohio

These system integrators updated their aircraft and engine designs and technologies; performed validation of their designs, utility, and cost analysis; and developed technology maturation roadmaps.

Additional funds were received by Northrop Grumman Corporation to conduct flight demonstration of direct sonic boom mitigation using a modified F-5E aircraft.

A special nose glove was designed for modification of aircraft to produce a shaped sonic boom profile with a lower impact at the ground level. Before the flight demonstration, tests done in wind tunnel validated the computed sonic boom signature predictions for the modified F-5E aircraft. A series of flight tests validated the predicted persistence of shaped sonic booms.

This program was very important because it demonstrated for the first time that an appropriately shaped aircraft can mitigate of sonic boom.

The experimental F-5E aircraft modified by Northrop Grumman Corporation (named F-5 Shaped Sonic Boom Demonstrator (SSBD)) is presented in **Figure 3** [23].

The theory was proven to work under practical design, fabrication, flight, and atmospheric conditions. Results of tests confirmed that shaping was successful in altering the sonic boom signature at the ground. Ground measurements matched predictions (flattop modified waveform relative to N-wave unmodified vehicle, **Figure 4**) [23]. In **Figure 4**, one can see that the "N" wave is no longer sharp in the case of shaped nose of F-5 SSBD (blue line) in comparison with the case of unmodified aircraft F-5E (red line). During this experiment, sonic boom reduction technology worked by achieving a shaped sonic boom, validating that shocks could be kept from coalescence all the way to the ground.

The image of modified aircraft from **Figure 3** shows at a glance the important drawbacks of this solution, affecting aerodynamic characteristics, frame's strength, weight, useful volume, and manufacturing cost of aircraft. An acceptable compromise is difficult to be found especially in the case of large passenger aircraft.

These drawbacks of shaping solution oriented the aircraft manufacturers to solution of supersonic aircraft with very thin fuselages. The first supersonic business jet is expected as to be Aerion AS2 which will be launched on market in 2023 (**Figure 5**) [24].

Main characteristics of this aircraft are [24]:

Supercruise: 1.4 Mach

Figure 3.
The F-5E aircraft modified by Northrop Grumman [23].

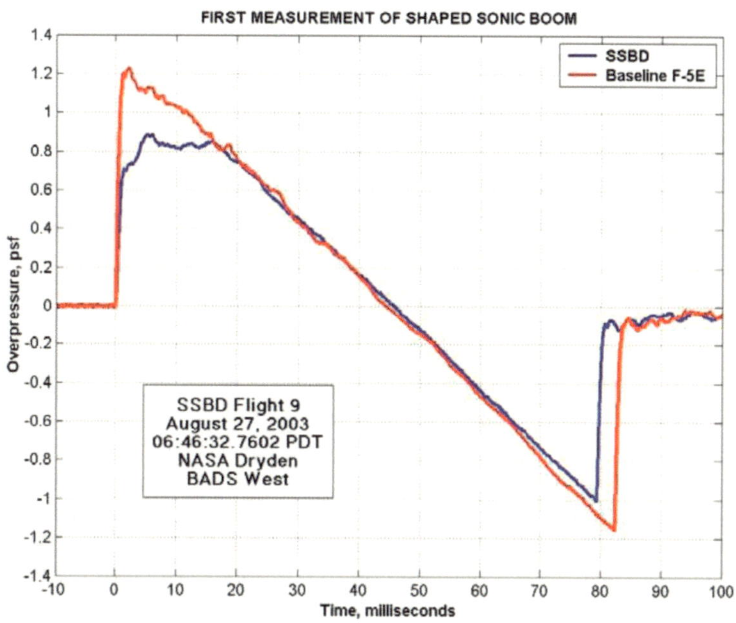

Figure 4.
First measurement of F-5E-shaped sonic boom aircraft modified by Northrop Grumman [23].

Boomless cruise: 1.1–1.2 Mach
Long range cruise: 0.95 Mach
Max. range, Mach 1.4: 7780 km
Max. range, Mach 0.95: 10,000 km
Wing area: 140 sq.m
Interior dimensions:
Height: 1.9 m
Width: 2.2 m
Cabin length: 9.1 m
Exterior dimensions:
Length: 51.8 m

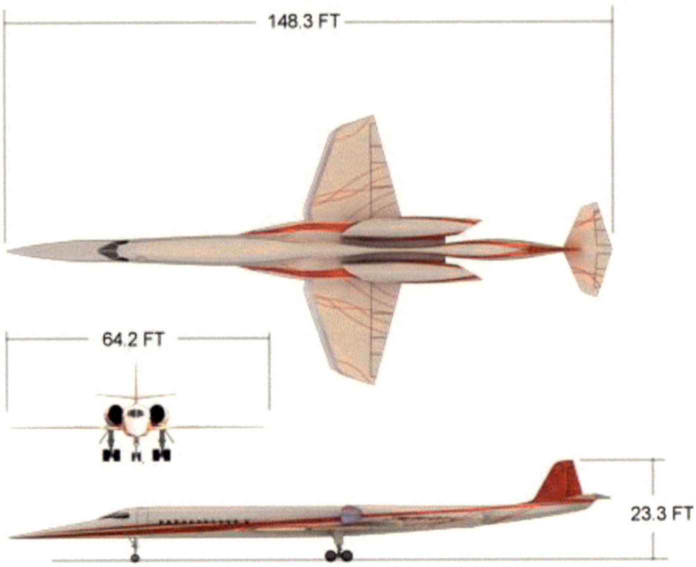

Figure 5.
Aerion supersonic business jet [24].

Wingspan: 23.5 m
Height: 6.7 m
Fuel quantity: 26,800 kg

Looking to the lengths of cabin (9.1 m) and aircraft (51.8 m), one can see at a glance one of the most important drawbacks of this solution: The space for passengers is extremely low due to the need of the aircraft fuselage to be very thin and long.

3. The theory of Sonics: A quick review

In 1918, the Romanian scientist George Constantinescu published *The Theory of Sonics* [25]. This book presents a new theory on the use of waves in the production, transport, and conversion of mechanical energy, as well as experimental validation. Constantinescu applied his theory to longitudinal waves of pressure propagating

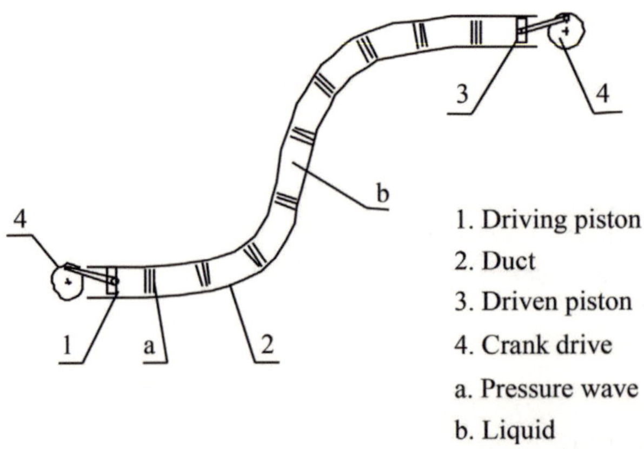

Figure 6.
Basic principle of the theory of Sonics [25].

through liquids, which fill metallic ducts. These ducts act as "wave guides" (see **Figure 6**). Piston, 1, oscillates in a sinusoidal manner and creates longitudinal waves of pressure, a. These waves propagate through liquid, b, which fills duct, 2, and actuates driven piston 3. Pistons 1 and 3 are going to oscillate with the same frequency. Crank drives, 4, assure the continuous motion of pistons. This method of power transmission relies on liquid compressibility. The phase difference between pistons 1 and 3 depends on the ratio of duct length and wavelength. If this ratio is an odd number, pistons 3 and 1 oscillate in opposition (i.e., the phase difference is equal to π). The amount of power that can be transmitted is proportional to the pressure of liquid within duct. Finally, George Constantinescu demonstrated that sonic waves act like alternative current and built many wave generators and sonic engines with power of tens of kW. Frequencies of sonic waves used for power transmitting can be from several tens to tens of thousands of Hz.

4. New solution for sonic boom mitigation

This new solution was proposed for the first time in a previous paper of authors [26]. It consists in dispersion of shock wave during its generation by an aircraft in supersonic flight having as a consequence extension of "N" wave (sonic boom) on a much larger area at ground level. In this way, the impact of sonic boom on community is much reduced.

This solution offers to aircraft designers the possibility to create supersonic aircraft with a larger space in fuselage and transportation of a higher number of passengers.

4.1 The bases of the new solution

The new proposed solution for sonic boom mitigation is based on the following observations:

1. The shock wave is a steady-state effect, which appears when the speed of aircraft is higher than the speed of sound in air.

2. For low values of Mach no. (M = 1, ... 1.8), a low variation of the semi-angle α of a wedge, which is placed in a supersonic stream produces a larger variation of shock wave angle, β.

The thickness of shock wave is extremely small. This thickness depends by Mach number as presented in **Figure 7** [27]. For this reason, when the shock wave hits the ground, a sudden increase of local air pressure is produced.

According to Observation 1, in normal circumstances, the shock wave cannot be eliminated because it is a physical effect governed by natural laws. However, if circumstances are changed, for example, the steady-state is substituted with a transient state; the effect of sonic boom on ground surface will be much reduced.

Taking as example the oblique shock wave created by a wedge having the semi-angle α (**Figure 8**), the semi-angle β of the shock wave is given by Eq. (1) [28]. Looking to Eq. (1), one can see that β is depending on the semi-angle α and the speed of aircraft given by the Mach number, M:

$$\cot\alpha = \tan\beta\left[\frac{(k+1)M^2}{2(M^2\sin^2\beta - 1)} - 1\right] \tag{1}$$

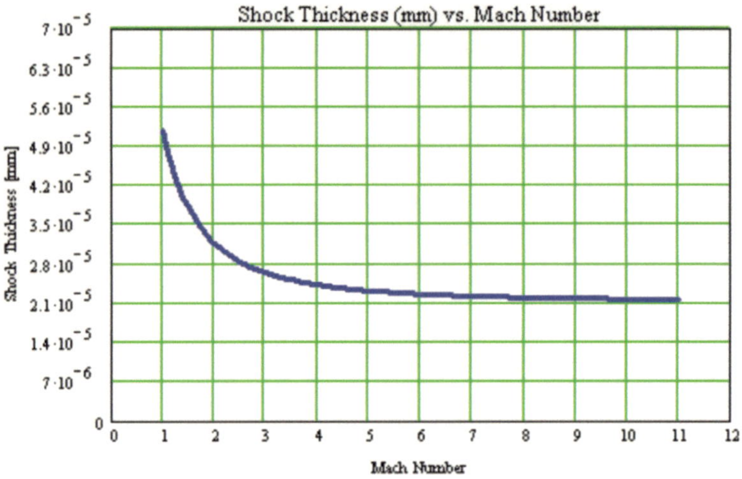

Figure 7.
The thickness of shock wave [27].

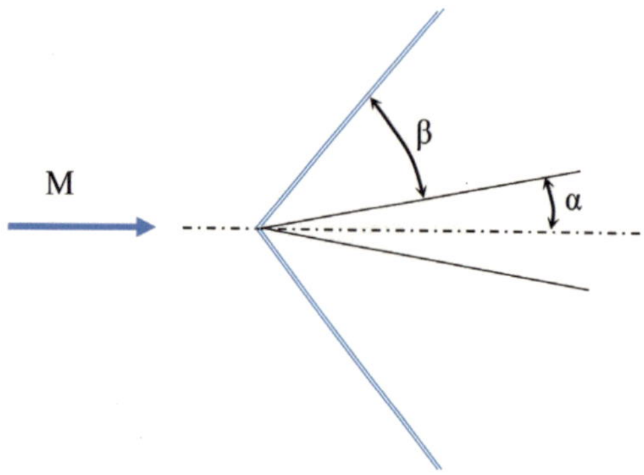

Figure 8.
The oblique shock wave.

A transient state could be produced in two ways:

a. Increasing and decreasing of aircraft speed (Mach number, M)

b. Increasing and decreasing rapidly the semi-angle α.

The first way (a) is impossible due to inertia. Really, it is obviously for everybody that the aircraft cannot be accelerated and decelerated rapidly because the thrust of engines cannot be increased and decreased rapidly.

The second way (b) is affordable if the supersonic aircraft is equipped with an equipment for dispersing of shock wave during flight over populated areas.

In this case the dispersion of shock wave, i.e., variation of angle β, is produced through periodical variation of semi-angle α of aircraft surfaces, which generate the shock waves, i.e., nose, wing leading edge (LE), and horizontal empennage LE.

During horizontal flying of a supersonic aircraft, its nose produces a conical shock wave, and the wing and horizontal empennage are producing oblique shock waves.

Therefore, three booms should be heard at ground level, but the second and the third booms are very close, and practically only two booms are heard.

During the travel of the three shock waves to ground, "N"-shaped wave is formed through coalescence hitting the ground as sonic boom. This "N" wave is composed by a high-pressure zone, where maximum pressure is $+P_0$ followed by a depression zone where minimum depression is $-P_0$.

4.2 Mechanical dispersion of shock wave using elastic membranes

In normal case, the shock wave thickness δ is extremely small as presented in **Figure 7**, and the footprint length d of the "N" wave at ground level is about two times larger the aircraft length.

For simplicity, assume an aircraft wing having the wing LE as a wedge, which can be continuously vibrated with a certain frequency, ν (**Figure 9**) [26].

Vibration of wing LE surface is done in this case by an elastic membrane, which is stretched over the wing LE. Between the wing and membrane, a thin layer of hydraulic liquid is introduced. When pressure pulses of a certain frequency ν are injected in liquid through perforations in wing LE, the membrane begins to vibrate with the same frequency ν. The pressure pulses can be produced by a sonic equipment as presented in **Figure 6**. In this case, the driven piston 3 from **Figure 6** is substituted by the elastic membrane.

For reaching of a high vibration amplitude, the injection frequency of pulses must coincide with the first resonance frequency of membrane. The resonance frequency of membrane depends on the value of stretching tension of that membrane over the wing LE.

In **Figure 9**, one can see that when semi-angle α increases, the shock wave semi-angle β increases, and the shock wave is dispersed on a larger area D > d. Due to dispersion, the thickness of shock wave at ground (S) is much larger than the thickness of the shock wave (δ) in the absence of vibration (S >> δ). Extension of shock wave on a larger area at ground level makes the maximum pressure $p_0 < < P_0$ and the impact of sonic boom on community to be much reduced.

According to observation 2, if the Mach number is between 1 and 1.8 (the case of the most supersonic business jet ongoing projects), a small variation of semi-angle α

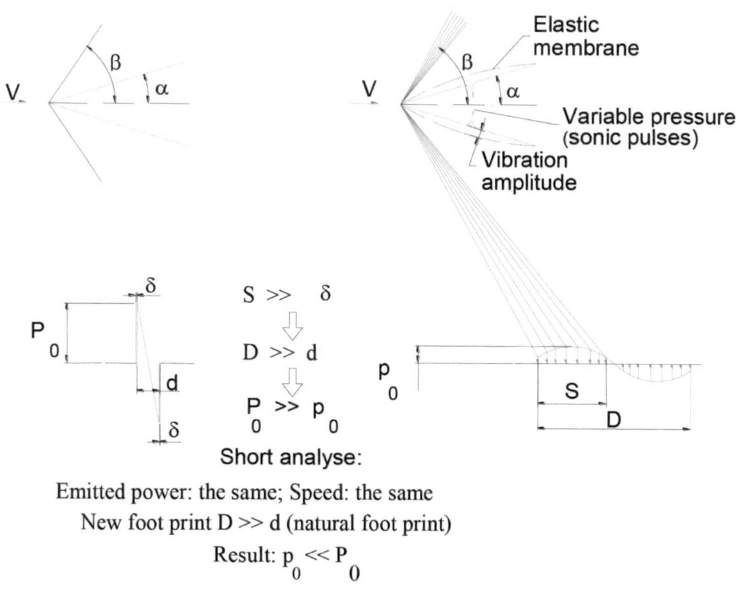

Figure 9.
Dispersion of shock wave by vibrating surfaces [26].

produces a large variation of semi-angle β. In the function of the Mach number, the variation of β can be even of several times larger than variation of α. This physical effect offers an important advantage: At the ground level, the dispersed shock wave extends on hundreds of meters. For example, if an aircraft is flying horizontally with speed M = 1.3, and the semi-angle of wing LE is α = 5°, the semi-angle β calculated with Eq. (1) is β = 59.96° [29]. If the semi-angle is α = 6°, the semi-angle β calculated with Eq. (1) is β = 63.46° [29]. So, if the variation of wedge semi-angle is Δα = 6° − 5° = 1°, the variation of semi-angle β is Δβ = 63.46° − 59.96° = 3.5°, i.e., much larger than Δα.

Assume M = 1.3, Δα = 1°, and Δβ = 3.5° (0.061 rad). If the aircraft is flying at height H = 15,000 m, shock wave dispersion is given by Eq. (2):

$$S = H \bullet \Delta\beta = 15000 \bullet 0.061 = 915\, m \tag{2}$$

If a supersonic aircraft has the length of 20 m, the natural ground footprint of the "N" wave is 40 m. One can easily see that through dispersion the footprint is enlarged about 23 times from 40 to 915 m.

However, even larger dispersion distances S can be obtained if through design the semi-angle α of wing LE is taken equally to α_{lim} for detaching of oblique shock wave. For a given supersonic cruise speed M, if the semi-angle α of wing LE is increased through vibration, only a little over α_{lim}, an extremely large variation of shock wave semi-angle β is produced. This is happening because when the semi-angle α is over α_{lim}, the oblique shock wave is detaching as presented in **Figure 10** [26].

The results of calculations using [29] are given in **Table 1**.

As it can be seen in **Table 1**, shock wave dispersion at the ground level is of thousands of meters.

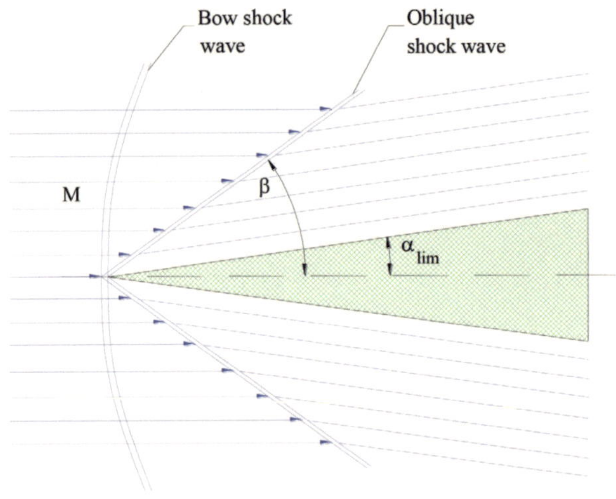

Figure 10.
Detaching of shock wave [26].

M	α_{lim} [°]	β_{before} [°]	β_{after} [°]	Δβ [°]	S [m]
1.300	6.650	68.59	90	21.41	5602
1.501	12.125	65.80	90	24.20	6332

Table 1.
Variation of β when α = α_{lim}.

For M = 1.501, if cone angle is α = 12.125°, a variation $\Delta\alpha$ = 0.025° theoretically produces transforming of oblique shock wave in a detached shock wave (bow shock wave).

Of course, keeping in control of such a process is a fine task, but it can be achieved if it is controlled by the aircraft onboard computer.

An important question is the following:

Which could be the most effective vibration frequency, ν? This is a difficult question. It is very clear that the effect of a low frequency, say 1 Hz, has no significant influence to shock wave dispersion because the vibration is too slow. Duration of a natural "N" wave is about 0.1 s. Probably, the period T of membrane oscillation should be smaller than 0.1 s, that is, the vibration frequency should be over ν = 1/T = 1/0.1s = 10 Hz.

At this time there is no theory regarding how much could be this frequency. For this reason, experiments are the next necessary step. Some experiments having an acceptable price are presented at point 5.

4.3 Mechanical dispersion of shock wave using elastic fairings

Obviously, applying elastic membranes on aircraft nose and wing LE implies a difficult technology. Instead of that design, a new one can be seen in **Figures 11** and **12** [30]. This time the membrane is substituted by an elastic fairing made of thin carbon fiber composite fixed by the aircraft nose or wing LE. When the pressure in the air manifold varies (e.g., sinusoidal variation with frequency ν = 10 Hz), local forces appear on elastic fairing determining its vibration. The variation of pressure must be equal to the resonance frequency of elastic fairing for obtaining the maximum vibration amplitude with a minimum pneumatic power.

Testing of such a solution has an acceptable price if the following two methods are applied.

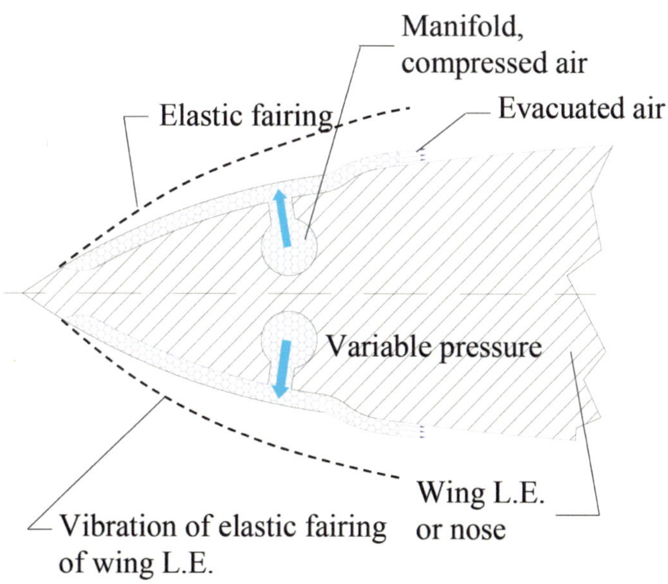

Figure 11.
Principal scheme of shock wave dispersion through vibration of elastic fairings induced by compressed air [30].

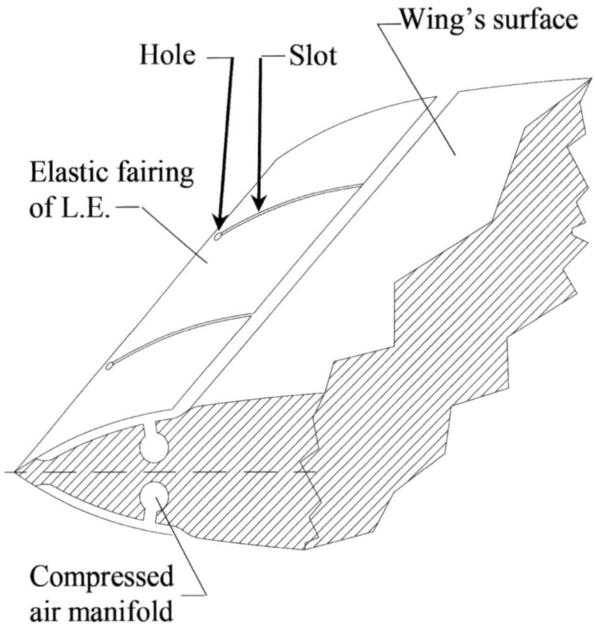

Figure 12.
View of a wing with elastic fairings at LE for dispersion of shock wave [30].

5. Proposed experimental tests for proving of concept

The effect of shock wave dispersion through mechanical vibrations can be tested in supersonic wind tunnel. At points 4.1 and 4.2, two test equipment are proposed.

The existing methods as schlieren photography are good for observing shock wave dispersion in supersonic wind tunnel.

5.1 The test equipment no. 1

At this equipment, a wedge composed of two elastic lamellae actuated by an electromagnet is used (see **Figure 13**) [26].

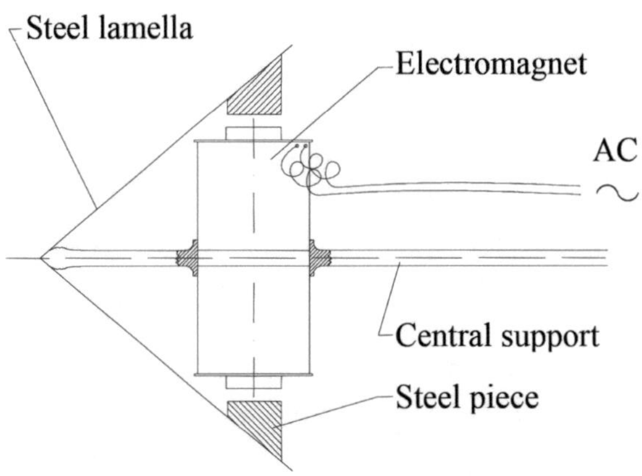

Figure 13.
The scheme of test equipment no. 1 [26].

The components of experimental equipment no. 1 are:

- Two steel lamellae having at their end steel pieces.
- Electromagnet that is fed by an alternative current (AC).
- Central support.

When electromagnet is powered with an alternative current at the frequency ν equal to the resonance frequency of lamella, the lamellae vibrate at the maximum amplitude. Vibration frequency can be changed if the mass of the two steel pieces is changed. When the weight of steel piece increases, the resonance frequency of lamella decreases and vice-versa. Another role of steel piece is to increase the attraction force of electromagnet on lamella.

Firstly, the shock wave is observed in the window of supersonic aerodynamic tunnel for various speeds when electromagnet is not actuated. The position of shock wave is schlieren photographed for various values of Mach number.

After that, the electromagnet is actuated by the AC having a frequency ν equally to the resonance frequency of lamella, and the shock wave is schlieren photographed for the same Mach number as before (when electromagnet was not actuated).

For every measurement, the shock wave should have variable taper and thickness, depending on vibration frequency and Mach number.

5.2 The test equipment no. 2

The test equipment no. 2 is more complex than test equipment no. 1. It should normally be used as a second step if good measurements are registered during using of equipment no. 1.

This equipment is presented in **Figure 14** [26]. The components of experimental equipment no. 2 are:

- Wedge simulating a cone or wing LE.
- Membrane made of elastic material.
- Hydraulic liquid.

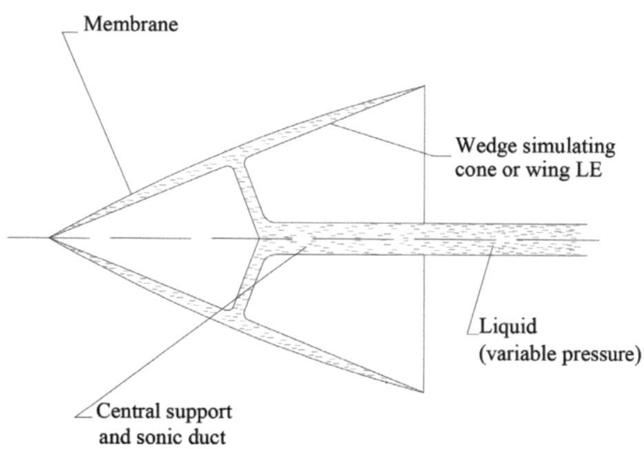

Figure 14.
Test equipment no. 2 [26].

A sonic generator sends pressure pulses to the main duct containing hydraulic liquid. The pressure pulses propagate with high speed to the liquid existent between membrane and wedge. As a result, membrane surface is bending and wedge angle increases with $\Delta\alpha$.

The experimenting procedure is similar to that presented at point 4.1: Firstly, the shock wave is observed for various speeds when sonic pulses through the main duct with hydraulic liquid are not present. Position of shock wave is schlieren photographed for various values of Mach number.

After that, sonic pulses are sent through the main duct with hydraulic liquid, and the shock wave is schlieren photographed for the same speeds as before.

For every measurement, the shock wave should have a variable taper and thickness depending on frequency and Mach number.

The experiments using the test equipment no. 2 are very useful because they simulate very close the real case on aircraft.

6. The supersonic European business and passenger aircraft

The European community intends to enter the competition for manufacturing of the future supersonic business and passenger aircraft. The future European supersonic aircraft could have a normal design except the nose and LE of wing and horizontal empennage. In a more sophisticated case, even the vertical empennage and the entry in engine admission device can be vibrated (**Figure 15** [26]). In the indicated areas, vibrating membranes or fairings should be mounted for shock wave dispersion.

7. Electrical solutions for dispersing of shock wave

Intense research is taking place in our days for dispersion of shock waves generated by aircraft through electrical means. It was observed that the so called plasma actuators consisting of high voltage electrodes (cathodes and anodes) have effect on airflow through air ionization. At present plasma actuators are researched both for noise reduction in the fan ducts of jet engines and for dispersion of shock wave for mitigation of its effects at ground level.

7.1 Using of plasma actuators for dispersion of shock wave

In some preliminary experiments, plasma actuators (**Figure 16**) were used for increasing of semi-angle β of conical shock wave (**Figure 16b**) [31].

When the potential difference between two electrodes (cathode, anode) (**Figure 16a**) is increasing progressively, the cone semi-angle β (**Figure 16b**) is increasing to a critical value β_{cr} when the shock wave becomes detached (**Figure 17**,

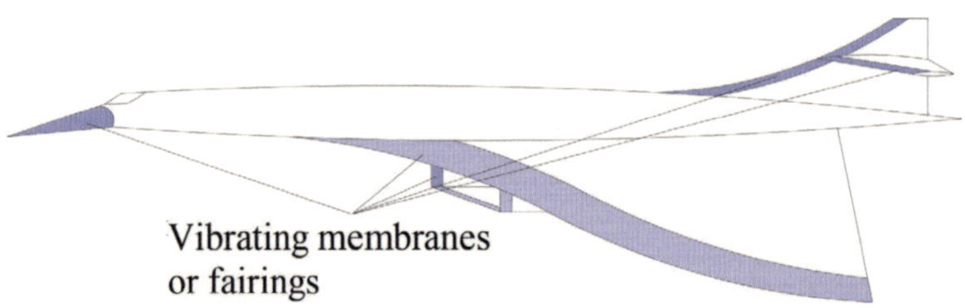

Figure 15.
A possible European supersonic business jet/passenger aircraft using vibrating surfaces for shock wave dispersion [26].

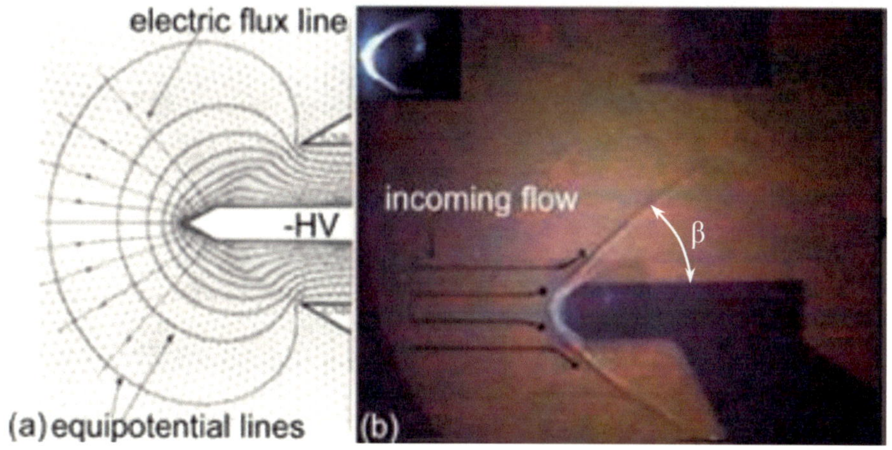

Figure 16.
Dispersing of shock wave using plasma actuators [31].

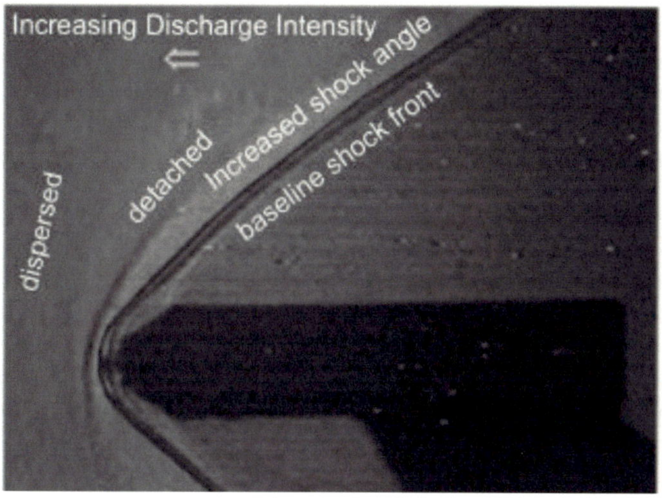

Figure 17.
Detaching of shock wave for a given potential [31].

Taylor-Maccoll theory). The potential difference between the two electrodes is of several thousands of volts.

In the presented case, positive ions are generated when atoms are losing an electron. An avalanche effect is taking place when electromagnetic energy is ionizing more atoms (this effect is visible as a blue light). Applying of such a solution seems to be difficult in the case of real aircraft because the electrical discharge can become thermal destroying in this way the electrode surfaces.

On the other hand, the aircraft nose and wing LE have a large area, and it is hard to believe that such a system, which was tested at low scale can be applied at the large scale of an aircraft.

7.2 A new possible solution for shock wave dispersing through injection of electrons in surrounding airflow through sharp electrodes

The new possible solution proposed in this paper is based on a massive injection of electrons through very fine and sharp electrodes in the upstream of air stream (in front of aircraft). This solution could be applied in future at the new supersonic jet called Concorde Mark 2 (**Figure 18** [32]).

The supersonic passenger aircraft Concorde Mark 2 filled in patent [33] by Astrium SAS and European Aeronautic Defense and Space Company would be capable to fly with 4023 km/h (1118 m/s) transporting 20 passengers or 3 tons of cargo on a distance of 8851 km. The duration of travel between London and New York would be of 1 hour.

The cathodes are sharp Wolfram needles placed along a rod, which is fixed in the tip of aircraft nose and along the wing LE (**Figure 18**). The anodes are thin copper sheets, which are fixed by the aircraft nose and pressure/suction sides of wing. Obviously, the anodes and cathodes are electrically insulated by the aircraft frame.

[Note: In **Figure 18**, the dimensions of cathodes are exaggerated for clarity. Actually, they have the dimension of a usual sewing needle].

The system works as follows:

A high potential electrical source (thousands of volts) is connected to the cathodes and anodes by means of an electrically insulated wire network. When the electrical high voltage source is connected to the wire network, a high number of electrons are released through the sharp tips of the cathodes.

[Note: This type of discharge differs by the type of discharge presented in Chapter 6.1 where positive ions are generated through loosing of electrons by atoms due to the primary electrons generated by cathode and accelerated by the potential difference between the cathode and anode].

The released electros are spread in the air stream without generating a significant number of ions because the distance between cathodes and anodes is much larger than in the case of plasma actuators.

The quantity of electrons injected in air stream is very high due to the high number of cathodes and their sharpness and the high potential applied. After detaching the sharp cathodes, the electrons move together with the oxygen and nitrogen molecules to the shock wave, which has the semi-angle β given by Eq. (1) (in the case of oblique shock wave). The injected electrons can be free among the oxygen and nitrogen molecules or can be temporary attached by a part of molecules generating in this way temporary negative molecules.

In this way, the shock wave will be composed of neutral nitrogen and oxygen molecules, free electrons and temporary negative molecules. The shock wave is

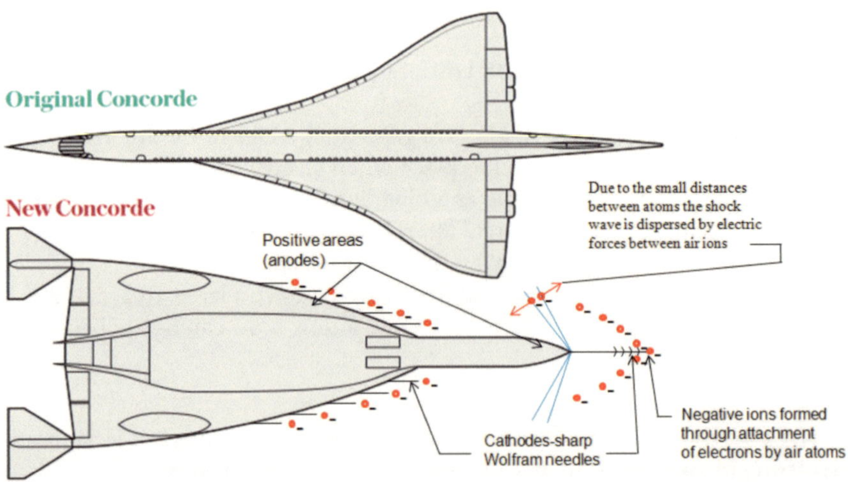

Figure 18.
New solution proposed for dispersing of shock wave through injection of electrons in surrounding airflow by sharp electrodes [16, 32, 33].

extremely thin (**Figure 7**, [27]). Inside the shock wave, due to the very small space, the density of electrons and temporary negative molecules is high. As a result, due to electrostatic repelling forces, the shock wave thickness must increase, and its impact at ground level will be mitigated.

After passing through the shock wave, the airstream is neutralized by the anodes placed on the aircraft nose and wing, which collect the electrons present in the air stream.

Due to the very high complexity of phenomena, it is risky to make theoretical predictions at this time. The best methodology is to do experiments in a supersonic wind tunnel using various configurations of electrodes connected at high potentials for observing shock wave shape. The shock wave should become weaker similar to the image presented in **Figure 17**. In that case, the significance will be that impact of sonic boom at ground level is mitigated in comparison with the normal case.

If the present system applied at Concorde Mark 2 will be the case, it should be activated in ascending and descending phase when the impact of sonic boom on community is maximum. When the aircraft is flying at very high heights or over the ocean, activation of system is not necessary.

8. Conclusions

The new solutions presented in this chapter use dispersion of shock wave through mechanical or electrical means. These solutions are alternatives for "shaping" solution or using of very thin fuselage.

Following the shock wave dispersion, the resulting sonic boom is spread on a much larger area at the ground level, as a consequence, the air in the 'N' shock wave is much smaller than in the normal case.

Low amplitude mechanical vibration of aircraft nose, wing LE, and horizontal empennage LE leads to shock wave dispersion.

A first technological possibility is vibrating a membrane, which is stretched over aircraft nose, wing LE, and horizontal empennage LE. In this case, the membranes are actuated by sonic pulses propagated through a hydraulic liquid.

A second solution is vibrating of elastic fairings placed over the aircraft nose, wing LE, and horizontal empennage LE.

Injection of electrons in front of aircraft cone/wing/empennage could be a productive technology for reduction of sonic boom impact on community in the case of supersonic/hypersonic passenger aircraft and business jets.

The cathode (negative electrode) is composed of multiple needles of Wolfram placed on a rod, which is fixed in the tip of aircraft cone or placed along the leading edges of wing and horizontal tail.

The anode (positive electrode) is composed of multiple copper plates glued by aircraft nose and wing suction/pressure sides.

The cathodes and anodes are electrically insulated by the aircraft frame.

The electrons released by the sharp cathodes in the airstream are free or can be attached by oxygen and nitrogen molecules forming temporary negative molecules.

When arriving in the shock wave, these temporary negative molecules and free electrons repel each other dispersing the shock wave. As a result, the impact of "N" shock wave at ground level will be much reduced.

Experiments should be initiated for evaluation of this possible effect. Voltages of many thousands of V should be used because the number of electrons injected in the air stream depends on the value of potential difference between the cathodes and anodes.

Acknowledgements

This represents original research of The Romanian Institute for Research and Development of Gas Turbines (COMOT) and Structural Management Solutions, both located in Bucharest, Romania.

The original solutions presented in this chapter were filed in patent.

Conflict of interest

Declaration: There is no conflict of interest related by scientific information presented in this Chapter.

Nomenclature

d	the length of natural foot print of sonic boom, m
D	the footprint length of the extended "N" wave of sonic boom when vibration is applied, m
H	the flight altitude of aircraft, m
M	Mach number, dimensionless
P_0	the maximum pressure of the natural "N" wave of sonic boom, Pa
p_0	the maximum pressure of the extended "N" wave of sonic boom when vibration is applied, Pa
S	the enlarged thickness of shock wave at ground level due to vibration, m
T	vibration period, s

Greek:

δ	the natural thickness of shock wave, m
ν	the vibration frequency of shock wave, s^{-1}

Author details

Constantin Sandu[1*], Radu-Constantin Sandu[2] and Cristian-Teodor Olariu[1]

1 Romanian Research and Development Institute for Gas Turbines-Comoti, Bucharest, Romania

2 Structural Management Solutions, Bucharest, Romania

*Address all correspondence to: constantin.sandu@comoti.ro

References

[1] Smithsonian National and Space Museum. Bell X-1 Glamorous Glennis [Internet]. Available from: https://airandspace.si.edu/collection-objects/bell-x-1 [Accessed: 11 January 2018]

[2] Walkden F. The shock pattern of a wing-body combination far from the flight path. Aeronautical Quarterly. 1958;**9**(2):164-194

[3] Parker MA. The sonic boom problem. Aerospace Engineering. 1968;**40**(8):30-38

[4] Zhilin YL. Effect of the aircraft configuration, its flight regime, and state of the atmosphere on the sonic boom intensity. Trudy TsAGI. 1967;(1094):14-22

[5] Jones JB. Lover bounds for sonic bangs in the far field. Aeronautical Quarterly. 1967;**18**(1):1-21

[6] Ferri A, Ismail I. Effects of lengthwise lift distribution on sonic boom of SST configurations. AIAA Journal. 1969;**7**(8):1538

[7] Caractéristiques Techniques du CONCORDE [Internet]. Available from: http://museedelta.free.fr/concorde/concorde_car.htm [Accessed: 11 January 2018]

[8] Seebass R, George AR. Sonic boom minimization. Journal of the Acoustical Society of America. 1972;**51**(2):686-694. DOI: 10.1121/1.1912902

[9] Miller DS, Carlson HW. On the application of heat and forces fields to the sonic boom minimization problem. Journal of Aircraft. 1971;**8**(8):657-662

[10] Zhilin LY. Sonic boom generated by a supersonic air liner. Trudy TsAGI. 1983;(1489):41-45

[11] Chirkashenko VF, Yudintsev NY. Development of a Technique for Measuring Sonic Boom Parameters in Supersonic Wind Tunnels. Preprint No. 6-83. Novosibirsk: Inst. Theor. Appl. Mech., Sib. Branch, Russian Acad. of Sci; 1983

[12] Henne PA, Howe DC, Wolz RR, Hancock JL Jr. Supersonic Aircraft with Spike for Controlling and Reducing Sonic Boom. US6698684B1/2002.01.30. Georgia: Gulfstream Aerospace Corporation

[13] Graham BM, Hamrick HC. Airplane Configuration Designed for the Simultaneous Reduction of Drag and Sonic Boom. US4114836A/1977-03-09

[14] Espy M. Sonic Boom Reduction. US3709446A/1969-05-09

[15] Scott R. Supersonic Aircraft Shock Wave Energy Recovery System. US08603013/1990-07-23

[16] Gerhardt HA, Kerswell JF, Priestley RT, Gibson BT. Laminar Supersonic Transport Aircraft. US5842666A/1997-02-21. Northrop Grumman Corporation

[17] Laflin K. Fuselage Design for Sonic Boom Suppression of Supersonic Aircraft. US20080105783A1/2006-11-03. Cessna Aircraft Company

[18] Lugg RH. Supersonic Aircraft with Shockwave Canceling Aerodynamic Configuration. US20110133021A1/2009-09-29. Hypermach Aerospace Industries Incorporated

[19] Smith H. A review of supersonic business jet design. Aeronautical Journal—New Series. 2007;**111**(1126):761-776. DOI: 10.1017/S0001924000001883

[20] Carlson HW, Raymond BL, Mack RJ. Application of Sonic-Boom Minimization Concepts in Supersonic

Transport Design, NASA TN D-7218 [Internet]. 1973. Available from: https://ntrs.nasa.gov/archive/nasa/casi.ntrs.nasa.gov/19790006829.pdf [Accessed: 04 November 2018]

[21] Darden CM. Sonic Boom Minimization with Nose Bluntness Relaxation, NASA TP-1348 [Internet]. 1979. Available from: https://ntrs.nasa.gov/archive/nasa/casi.ntrs.nasa.gov/19790006829.pdf [Accessed: 04 November 2018]

[22] US Defense Advanced Research Projects Agency. Quiet Supersonic Platform Phase II Contractors Selected [Internet]. 2002. Available from: http://www.defense-aerospace.com/articles-view/release/3/9835/lockheed,-northrop-win-qsp-awards-(apr.-30).html [Accessed: 03-04 November 2018]

[23] Pawlowsky JW, Graham DH, Boccadoro CH, Coen PG, Maglieri DJ. Origins and Overview of Shaped Sonic Boom Demonstration Program, 43rd AIAA Aerospace Sciences Meeting and Exhibit; 10-13 January 2005; Reno, Nevada, Paper No. AIAA2005-5

[24] Aerion is Creating a New Era of Sustainable Supersonic Flight [Internet]. Available from: https://www.aerionsupersonic.com/ [Accessed: 03 November 2018]

[25] Constantinescu G. Teoria Sonicitatii. 2nd ed. Bucuresti: Editura Academiei Romane; 1985

[26] Sandu C, Brasoveanu D. New solution for sonic boom mitigation. Concept and testing methodology. Application at European supersonic business jet. In: Proceedings of the 33-rd Caius Iacob Conference-Fluid Mechanics and Technical Applications; 29-30 September 2011; Bucharest, Romania. 2011

[27] What is a Shock Wave [Internet]. Available from: http://www.aerorocket.com/Nozzle/Validate/Tshock/Shock.html [Accessed: 10 November 2018]

[28] Carafoli E. Aerodinamica Vitezelor Mari. 2nd ed. Bucuresti: Editura Academiei Republicii Populare Romane; 1957

[29] Compressible Aerodynamics Calculator [Internet]. Available from: http://www.dept.aoe.vt.edu/~devenpor/aoe3114/calc.html [Accessed: 10 November 2018]

[30] Sandu C, Brasoveanu D. Too large air stream deviation passenger spaceplanes and airplanes that have variable configuration for sonic boom reduction. In: Proceedings of the 6-th CEAS2017 Conference; 16-20 October, 2017; Bucharest, Romania

[31] Kuo SP. Air plasma mitigation of shock wave. Advances in Aerospace Science and Technology. 2016;**1**:59-69

[32] Concorde Mark 2. Airbus Files Plans for New Supersonic Jet [Internet]. Available from: https://www.telegraph.co.uk/finance/newsbysector/industry/engineering/11782446/Concorde-Mark-2-Airbus-files-plans-for-new-supersonic-jet.html [Accessed: 12 November 2018]

[33] Prampolini M, Coraboeuf I. Ultra-rapid Air-Vehicle and Related Method for Aerial Locomotion. Patent No. US 9,079, 661 B2; 2015

Chapter 7

A Mathematical Model of Noise Pollution in Streets of Tehran near IKIA Airport

Amir Esmael Forouhid

Abstract

This chapter attempts to investigate noise pollution in the street of Tehran near IKIA airport considering the population growth and large contribution of ground vehicles to noise pollution. The operation of airports results in environmental impact associated with high levels of different sources of noise. If there is to be growth in aviation, the environmental impacts of aviation must be mitigated. In this chapter, a model for the noise pollution near IKIA airport in Tehran has been calibrated with the use of a noise forecasting software. The study areas consisted of high traffic areas of Tajrish Street near the airport. In a field study, the noise level was measured via sound meter and the noise map was generated based on geo-statistical methods via GIS software. For this purpose, the factors influencing noise level (e.g. traffic, road width, slope, and residential or administrative-commercial land use) were surveyed and recorded for each point and their local and time dependencies were computed via SPSS. According to the results obtained from 324 survey points, the highest noise levels belonged to Tajrish area (69 dB). It is higher than standard noise level for residential and commercial and in conclusion, suggestions for reducing it are given.

Keywords: noise pollution, GIS, SPSS, regression, airport

1. Introduction

Public pressure led to the introduction of many different types of constraint at an increasing number of airports in an effort to keep both annoyance and complaints to a minimum. Noise pollution surrounding IKIA airport is a growing concern in Tehran.

The computer simulation of the noise exposure level that we use at IKIA airport and its surrounding areas is conducted using the noise exposure forcasting modeling, The Noise Exposure Forecasting modeling, as used in the current study, computes noise exposure levels. The noise metric computed by the model is the annual average Day–Night Sound Level. The widely-used DNL metric is known to be highly correlated with community annoyance and is associated with a variety of land use guidelines that suggest where incompatibilities are expected to exist between the noise environment and various human activities [1].

Noise is an adverse factor in the living environments of today's communities. This type of pollution has drawn attention to itself in the three recent decades, being

IntechOpen

a major problem in larger cities and seen as one of the significant environmental problems which is on the rise due to an array of factors including increased population density, motor vehicles, industrial activities in the proximity of urban areas and construction activities. Above-standard noise levels negatively affect all living beings and are therefore classified as environmental pollution. Research shows that both short- and long-term exposure to noise pollution weakens hearing, increases blood pressure leading to cardiovascular disorders, causes sleep and mood disorders, and changes behavior patterns [2]. Basner et al. studied the auditory and non-auditory influences of noise on cardiac diseases and neural disorders [3]. Therefore studying noise pollution and generating noise level maps for metropolises such as Tehran is of significant importance. Sound measuring in each certain area would take a considerable time and cost [4] because the level of noise emitted by vehicles in streets varies depending on traffic condition and such variation needs to be considered in measurements.

Another way used to measure noise pollution in a certain area is employing noise pollution models, which indeed is a mathematical technique. Mostly traffic-based parameters, which are very diverse, are used as the model's inputs. There are many factors that may affect sound emission in the space; hence, numerous models have been developed in this area so far. Most of them are physical factors which cover sound properties. As a result, measuring and examining such parameters is very difficult and complicated. However, other parameters including traffic-related ones such as velocity of vehicles and traffic flow are measured very easily [4].

Many studies have focused on noise pollution with different methods. Bilasco et al. proposed an information system model to identify areas exposed to noise pollution. Their model generated a noise map using sound measurements, building heights, land uses, digital land altitude model, and wind speed and direction in GIS software [5]. Subramani et al. analyzed noise pollution at different crossroads. They first performed a time analysis on sound data in different time periods at different crossroads, thus generating the noise map via GIS [6]. In another work, Mendal et al. assessed and analyzed noise pollution in Kolkata, India at the time of a festival [7]. Abbaspour et al. performed a hierarchical analysis of noise pollution in a region of Tehran [8]. Investigating the research performed on noise pollution analysis indicates that for noise pollution studies, GIS system is highly capable in generating noise maps and data analysis. The most important influential factors on the subject are traffic, width and type of roads, urban land use, green space, and slope. Traffic is important because increased traffic means increased vehicles, therefore increased noise. Road width also influences the capacity for holding vehicles. Road slope influences vehicle speeds, with higher speeds generating higher levels of noise. Land use influences generated noise by changing the population density and commenting levels.

2. Research method

The paper studied Tehran, Iran. The study areas consisted of Tajrish Sq. (region 1) street of Tehran. The study areas were selected based on their traffic and urban importance. Measurements were performed during 7–8 am on August 5–8, 2018. The survey measured sound levels, road slope, road width, traffic, and land use (residential, commercial, administrative, and green space) that are presented in the following sections.

The computer simulation of the noise exposure level that use at IKIA airport and its surrounding areas is conducted using the noise exposure forecasting modeling, The noise exposure forecasting modeling, as used in the current study, computes noise exposure levels. The noise metric computed by the model is the annual average Day–Night Sound Level. The widely-used DNL metric is known to be highly correlated with community annoyance and is associated with a variety of land use guidelines that suggest where incompatibilities are expected to exist between the noise environments and various human activities. Data input to the NEF includes runway coordinates, flight tracks, flight operations and types of aircraft. NEF computes the overall noise exposure at points on the ground around the airport. Data was modeled for a period of one year. The use of NEF in computer-based noise modeling not only gives the noise exposure levels based on the current flight operations, but also allows for the prediction of future noise levels due to a projected increase in flight operations. This is especially useful for a rapidly growing city like Tehran and IKIA airport.

The data and assumptions used for leading such a study are presented and detailed below.

Data summary:

- Distance to city center: 40 km
- Airport site area: 13,500 Ha
- Airport reference point coordinates:
    - Latitude: 35°24′58″N;
    - Longitude: 051°09′08″E;
    - Elevation: 1007 m.

Specifications for all planned runways are summed up below:

- Length of scheduled runways is 4200 m;
- Width of runways is 60 m with 15 m shoulders;
- Distance between parallel runways will be 400 m;

According to IKIA existing Airport Information Publication (AIP), the procedures for outbound aircraft are mainly concentrated on seven radials originating from IMAM KHOMAINI VOR/DME or destination to KAHRIZAK NDB. These exit radials were used to model the flight tracks followed by the various aircraft that will be accommodated at IKIA airport. Straight-in approaches were assumed on each runway.

This paper uses the noise exposure forcasting because its availability. The NEF consists of a map of the noise contours plotted over the airport layout at each time period. Noise contours for NEF 30, 35, 40 and 45 noise levels are shown on the map.

In this software, NEF+35 = DECIBEL and for discussion, should change the numbers of NEF to decibels. In the chart below, define the steps for modeling the noise of the airport .

It does this first by making the runways shown in **Figure 1** and then with the data from the runway, start the software.

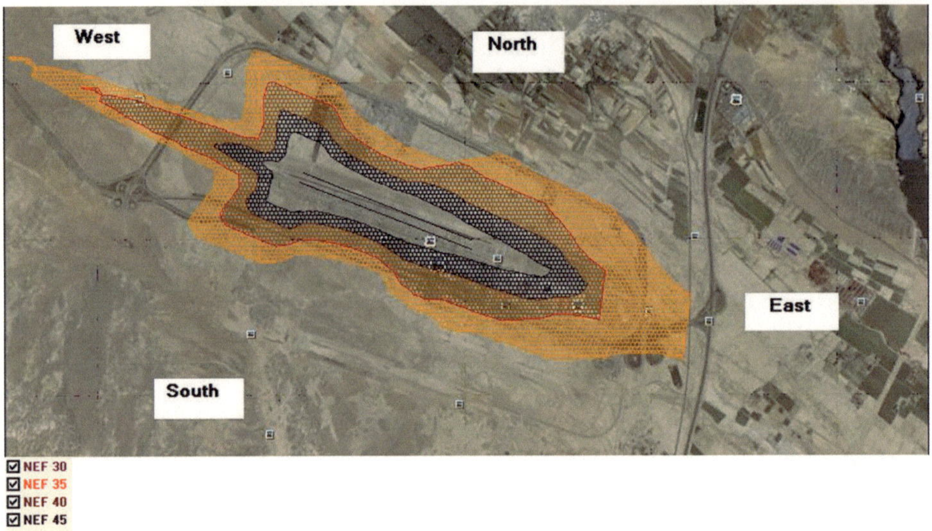

Figure 1.
Contours of the noise around the airport. Invert the output of software to decibels: Black contour: 80 dB; Brown contour: 75 dB; red contour: 70 dB; and at last contour 65 dB.

Night (10 pm–7 am) Unit in dB	Day (7 am–10 pm) Unit in dB	Type of region
45	55	Residential region
50	60	Residential-commercial region
55	65	Commercial region
60	70	Residential-industry region
65	75	Industry region

Table 1.
Standard for noise values [9].

Standard suggestion of the environmental organization of Iran for noise is shown in **Table 1**. From the results of contour map and this table, divide four the regions and in **Figure 1** the result is shown.

With comparison the noise map from the noise exposure forecast modeling with the ICAO land use recommendations in **Table 1**, and knowing that in the airport, we also have noise pollution from numerous vehicles and factories that may develop in the near future, should have a master plan for decreasing the noise of the airport, should do it first at the origin of it and then by barriers with a suitable plan for building near IKIA. The next section suggests some recommendations that may be used for the airport.

2.1 Influential parameters

2.1.1 Slope

Road slope greatly influences commuting, traffic arrangement, speed, and driving patterns. Noise pollution varies between roads with different slopes. The mean slope of the study areas was 0–3%.

2.1.2 Land use

The National Cartographic Center's 1:2000 maps, field studies, and Google Maps™ were used to calculate land use (residential, commercial, administrative, and natural ground).

2.1.3 Traffic

Considering the direct influence of urban traffic on the noise level, the authors measured raw data for the number of vehicles, survey time, and other data at each point in order to calculate the traffic of survey points via extra processing. The correlation between traffic level and noise level made this data essential. A reasonable high-traffic hour at the peak mounting traffic (7–8 am) was selected to measure peak traffic. The vehicle data, converted to their equivalent according to saloon vehicles, were calculated using Eq. (1) [10].

$$V_a = V \times C_f \tag{1}$$

where V_a is traffic volume in a complete period, V is the calculated traffic during the measurement time in a period, C_f is the correction factor of the measurement.

C_f is calculated via Eq. (2)

$$C_f = \frac{T_c}{T_c - T_s} \tag{2}$$

where T_c is the complete measurement period (min), T_s is the short stoppage time (min).

2.1.4 Road width

Due to high traffic and impossibility of direct measurements, road widths were estimated base on the number of lanes (every 3–3.65 m) and matching them with Google Earth maps and the mean value for multiple sections of each road.

2.2 Generating the noise level map

The sound data were collected using the device TES Sound Level Meter 1353H (calibrated by a qualified company) measurements were performed in the mornings at 15–20 m intervals. The longitude and latitude of each measurement point were recorded via a Garmin GPS device.

If during measurements, a vehicle with very high noise levels (bus, heavy truck, etc.) passed nearby the measuring device at low speeds or stopped, the authors attempted to remove its effect from measurements as it would introduce abnormal variations in measurements leading to statistical errors.

Data analysis was performed via ArcGIS 10.4.1. Raster polygon layers of sound data for the four studied areas are presented in **Figure 2**.

The data model:

After testing the relationship between noise level variable and different combinations of independent variables, the best model was selected using Eq. (3).

$$L_{eq\,(m)} = 70.554 + 0.002 \times Traffic - 0.078 \times Residential \tag{3}$$

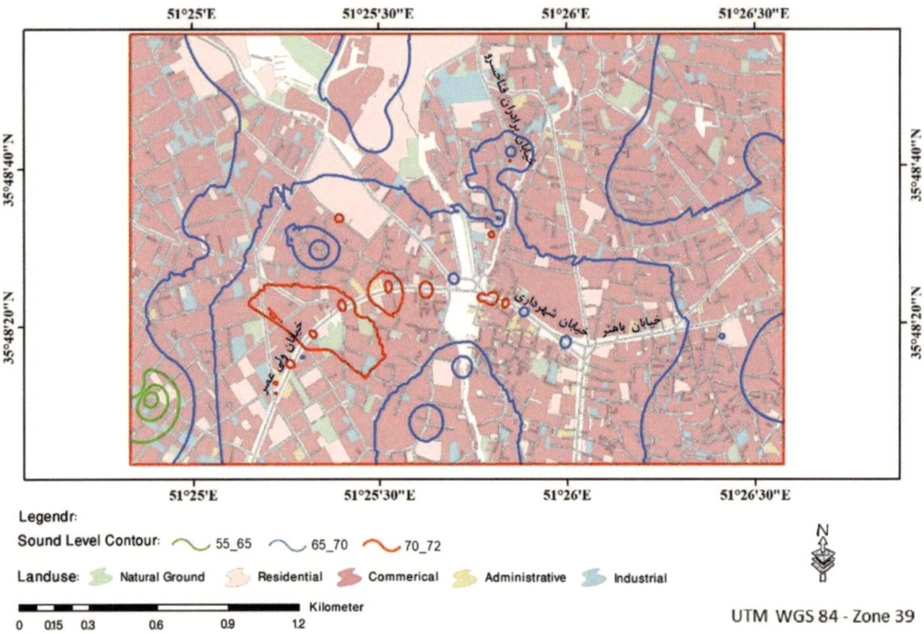

Figure 2.
Noise level map of Tajrish.

where $Leq_{(m)}$ is the noise level (dB), $Traffic_m$ is the vehicle traffic, and *Residential* denotes the percentage of residential land use.

As shown, noise level is related to the traffic volume and residential land use independent variables. There is a positive linear relationship between noise level and vehicle volume, indicating that higher vehicle volumes resulted in increased noise. The relationship predicts that one vehicle per hour increase in vehicle traffic volume will increase noise level by 0.002 dB. There is a negative, linear relationship between noise level and percentage of residential land use, indicating that 1% increase in residential land use will increase noise level by 0.078 dB. T**ables 2** and **3** show the data for observation times, goodness of fit indices (R^2, R^2_{Adj}), mean square error (MSE), results of t and F tests, and their respective significance, the estimated parameters (coefficients of independent variables) and their confidence interval, and the variance analysis table and F statistical value. The following section provides an analysis for each table.

2.2.1 Normality of the residuals

Figure 3 shows residuals according to their frequencies, representing a relatively normal distribution.

2.2.2 Linearly or nonlinearity of the relationships between dependent and independent variables

The linearity analysis was performed using a graph separating dependent and independent variables. According to **Figures 4** and **5**, the maximum value of goodness of fit index for the traffic and noise level relationship was 0.64, followed by 0.489 for the percentage of residential land use. Road width ranked third however it could simultaneously be used in the regression model due to very high correlation with the traffic variable.

	Traffic	sound Level	Slope	Residential	Commercial	Administrative	Natural Ground	Road Width
Traffic	1							
Sound Level	0.6931	1						
Slope	−0.5235	−0.4349	1					
Residential	−0.6140	−0.6486	0.6459	1				
Commercial	0.1575	−0.1258	−0.1251	0.0715	1			
Administrative	−0.6536	−0.3095	0.0749	0.0175	0.0557	1		
Natural Ground	0.5788	0.5876	−0.2807	−0.7782	−0.5012	−0.3898	1	
Road Width	0.8285	0.6242	−0.5469	−0.5731	−0.0680	−0.6015	0.7075	1

Table 2.
Variable correlations.

Model	R	R^2	Adjusted R^2	Std. error of the estimate	Change statistics					Durbin-Watson
					R^2 change	F change	df1	df2	Sig. F Change	
1	0.800	0.640	0.638	2.8233	0.641	497.107	1	280	0.000	
2	0.826	0.682	0.680	2.6577	0.042	36.980	1	279	0.000	1.936

Table 3.
Model summary.

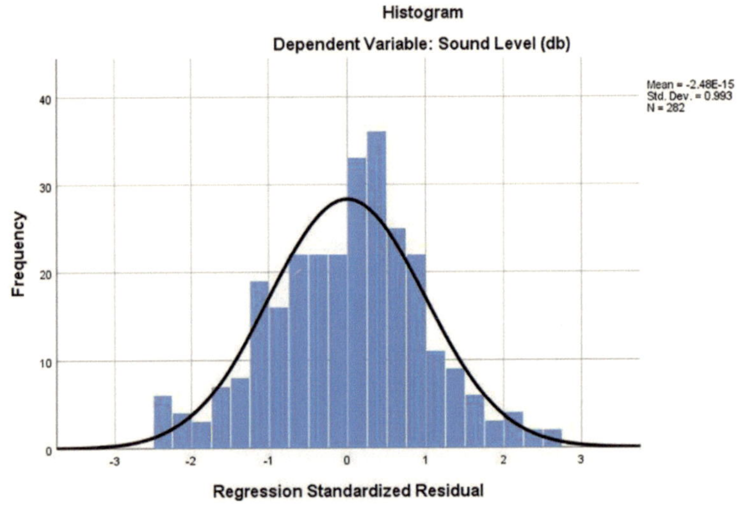

Figure 3.
The frequency of the regression model residuals.

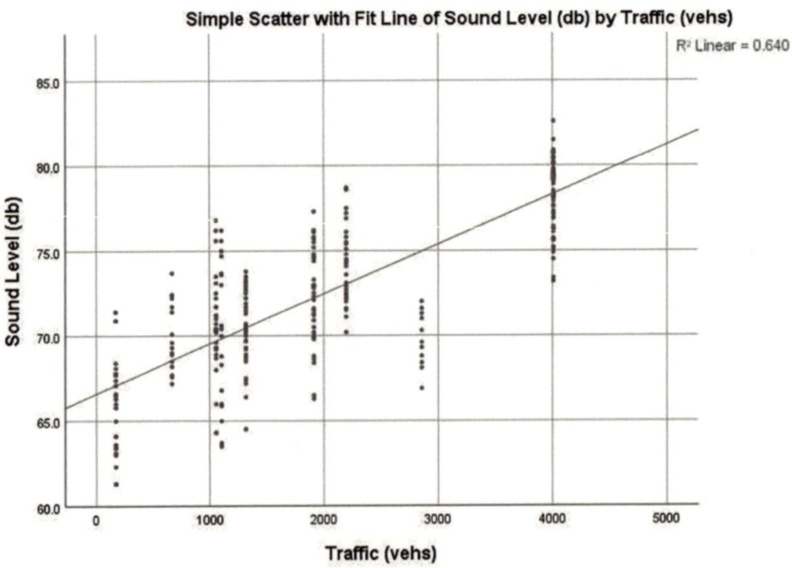

Figure 4.
Noise level and vehicle volume relationship.

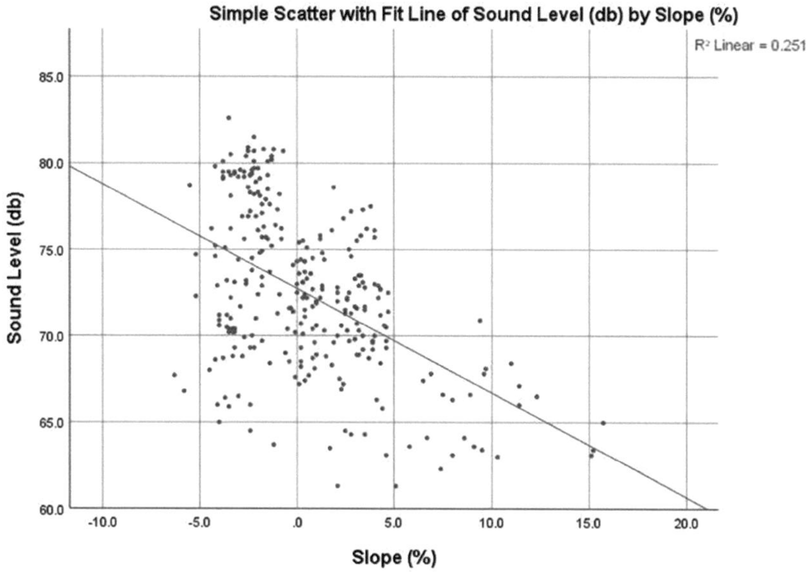

Figure 5.
Noise level and road slope relationship.

3. Conclusion

The negative impact of aircraft noise, in particular around airports, is increasing. More and more people suffer not only from annoyance, but recent studies indicate that intermediate and high noise levels also contribute to physiological and psychological effects that in extreme cases can cause severe health problems. The aircraft industry has launched an ambitious plan for the next 15 years to reduce the noise emission levels from aircraft by as much as 20 dB. Even if this goal can be reached, reduced noise emission levels for new aircraft will have little or no influence on the total noise situation around airports in future. This is due to a slow renewal rate for aircraft combined with an increase in passenger volume.

In order to stay competitive and to cope with an increasing number of neighborhood complaints and noise-impact related constraints, airport owners will have to look for novel solutions to reduce noise emission levels.

The International Civil Aviation Organization (ICAO) has defined a four-point "balanced approach" that includes:

Reduction of noise at source;

For improving this method airports authorities should develop and buy new aircrafts that have less noise such as boeing 757 instead of boeing 727 and etc.

Land-use planning;

The results indicate the critical significance of urban traffic in noise pollution, as by a large difference it had the highest contribution to noise level, followed by green space, administrative, and commercial land use; road width, and road slope.

Commercial and business land uses generated the highest noise pollutions. With their high commuting levels and passenger traffic, malls and commercial centers produce high noise levels, especially at certain hours in the morning, resulting in higher noise and environmental pollutions compared to natural ground or residential areas. Sound levels above 70 dB irritate humans.

For reduced noise pollution in Tehran and generally all urban areas, it is recommended to promote good driving behaviors and vehicle technical control for their sound level as well as implementing sound barriers for preventing the sound leaking into residential areas. Further, it is recommended that for future roads or revamping the existing ones, more lanes be implemented to produce wider roads, prevent the construction of tall buildings on the sided of main roads, and maintaining a standard distance between buildings and main roads, freeways, and other motorways.

The negative impact of aircraft noise, in particular around airports, is increasing. More and more people suffer not only from annoyance, but recent studies indicate that intermediate and high noise levels also contribute to physiological and psychological effects that in extreme cases can cause severe health problems. The aircraft industry has launched an ambitious plan for the coming 15 years to reduce the noise emission levels from aircraft by as much as 20 dB [1].

Other strategies for reducing noise pollution in urban areas include designating suitable locations for land uses in comprehensive and development plans, use of standard, low-noise vetches, imposing limitations on the passage of automobiles and motorcycles, imposing speed limits, improving traffic behaviors and extending public transport. Sound barriers around motorways and the use of sound-absorbent materials in commercial and residential buildings or natural ground near residential areas or roads will greatly reduce noise pollution levels. In addition, proper city-wide planning requires establishing sufficient noise-pollution measurement stations and sound level maps for different urban regions and land uses.

Author details

Amir Esmael Forouhid
Department of Civil Engineering, Parand Branch, Islamic Azad University, Parand, Iran

*Address all correspondence to: amiresmaelf@yahoo.com

References

[1] Forouhid AE, Mansour Khaki A, Yazdan Panah M. Noise Pollution Associated with the Operations of the IKIA Airport. Publisher: Environmental health perspectives. 16th Air Transport Research Society; 2012

[2] Passchier-Vermeer W, Passchier WF. Noise exposure and public health. Environmental Health Perspectives. 2000;**108**(Suppl 1):123-131

[3] Basner M, Babisch W, Davis A, Brink M, Clark C, Janssen S, et al. Auditory and non-auditory effects of noise on health. The Lancet. 2014;**383**(9925):1325-1332

[4] Golmohammadi R, Abbaspour M, Nassiri P, Mahjub H. Road traffic noise model. Journal of Research in Health Sciences. 2007;**28**:13-17

[5] Bilaşco Ş, Govor C, Roşca S, Vescan I, Filip S, Fodorean I. GIS model for identifying urban areas vulnerable to noise pollution: Case study. Frontiers of Earth Science. 2017;**11**(2):214-228

[6] Subramani T, Sounder S. A case study and analysis of noise pollution for Chennai using GIS. International Journal of Emerging Trends & Technology in Computer Science (IJETTCS). 2016;**5**(3):125-134

[7] Mandal PK, Bandyopadhyay A. Case study from West Bengal, India: Management of noise levels during festival time in Kolkata and Howrah Municipal Corporation Areas from 2002 to 2011. Environmental Quality Management. 2014;**23**:13-32

[8] Abbaspour M. Enviromenntal Engineering. Tehran: Islamic Azad University Publication, Scientific Publishing Center; 1992 (In Persian)

[9] International Civil Aviation Organization (ICAO), International Standards and Recommended Practices. Environmental Protection, Annex 16 to the Convention on International Civil Aviation, Volume I: Aircraft Noise. Air Transport Research Society. 1993

[10] Zoghi H, Haj Ali M, Jafari Haghighat Pour P. Traffic Engineering (Analytical-Applied). Tehran: Kian Publication; 2015 (In Persian)

Section 3

Emissions Mitigation Strategies

Chapter 8

Increasing the Emission Mitigation Potential by Employing an Economically Optimised Transport Aircraft Retirement Strategy

Oluwaferanmi Oguntona

Abstract

This study investigates the emission mitigation potential of retiring passenger aircraft economically at the global fleet level. In an integrated model of the air transport system, fleet turnover aspects of the global passenger aircraft fleet are defined using aircraft lifetime direct operating costs. Two fleet renewal strategies are compared in the study. The growth strategy (the baseline scenario) prioritises aircraft allocation for serving demand growth and replacing aircraft retired at the end of their design lives, before replacing those that are retired because of their operating cost disadvantage. Conversely, the replacement strategy allocates global aircraft production capacity first for replacing aircraft that are retired based on their operating cost disadvantage and those retired on reaching their design life limit, before serving growth in air travel demand. Results show that in year 2024, emission savings of three percent were achieved at the global fleet level using the Replacement Strategy, when compared to the baseline. Afterwards, due to the unavailability of newer efficient aircraft, emission savings diminish to two percent (around 40 million tonnes of CO_2). This research is useful to aviation stakeholders in having an overview of expected emission savings of the proposed strategic measure.

Keywords: fleet renewal strategies, aircraft direct operating cost, aviation emissions, emission mitigation, aircraft retirement

1. Introduction

There were around 28,000 commercial aircraft as of 2017, and out of these, 22,337 passenger jet aircraft have operations that yielded around 859 million tonnes of CO_2 [1–3]. Based on forecast of growth in air traffic, aviation emissions are also expected to grow in the long term. For example, Japan Aircraft Development Corporation [3] claimed that approximately 33,500 aircraft would be added to the global fleet over the next 20 years. The International Civil Aviation Organization [4] also claimed that a five percent growth in air traffic, compared to the projected one to two percent annual increase in aircraft fuel efficiency, would lead to the growth in emissions.

According to the International Air Transport Association [5], the air transport industry developed and has restated commitment to its aviation emission mitigation targets as follows:

i. An average improvement in fuel efficiency of 1.5% per year from 2009 to 2020

ii. A cap on net aviation CO_2 emissions from 2020 (carbon-neutral growth)

iii. A reduction in net aviation CO_2 emissions of 50% by 2050, relative to 2005 levels

A combination of different measures (technology, operations, infrastructure, economic and additional technologies and biofuels) would help achieve the 2050 target of reducing net aviation CO_2 emissions by 50%, relative to 2005 levels [6]. Thus, this study proposes a strategic measure of mitigating aviation emissions by giving priority to retiring passenger aircraft at the global fleet level when they are evaluated to have an operating cost disadvantage in comparison with other aircraft. The emission mitigation potential (EMP) of the measure is also evaluated.

The next section gives an overview of measures already proposed in different studies for mitigating aviation emissions. Section III describes the methodological approach used in the study. Subsequently, two fleet renewal strategies (*Replacement Strategy* and *Growth Strategy*) are presented before analysing the potential of using the *Replacement Strategy* over the *Growth Strategy*.

2. Review of measures for mitigating aviation emissions

The process of fleet development is majorly driven by the satisfaction of demand, subject to objective conditions. These conditions include profit maximisation or return on investment, minimising operating costs especially when faced with major maintenance events, environmental constraints effected as governmental restrictions and exogenous market dynamics like aircraft demand and supply and fuel prices [7]. Furthermore, before an airline decides to add aircraft to the fleet, an evaluation period is chosen to ensure that the aircraft introduced to the global fleet either to fill capacity gap or as replacement aircraft still gives a unit cost advantage by the end of the evaluation period. Wensveen [8] claimed a planning horizon of 10 years for a typical fleet planning model, while Clark [9] and Belobaba [10] stated possible periods between 6 and 12 and 10–15 years, respectively, for the macro-approach to fleet planning.

Extensive work has been done in proposing emission mitigation measures that can be applied at the aircraft fleet level, using various fleet development tools [11–21]. A review of these studies was done by Oguntona [22]. **Table 1** summarises the measures proposed in these studies and categorises them under the broad groups of measures for mitigating aviation emissions.

Although these studies have covered essential aspects of measures proposed by the industry, there has been no consideration of prioritising aircraft economic retirement as an emission mitigation measure.

3. Method

Fleet system dynamics model (FSDM), an integrated modelling environment (IME), was developed by Randt [23] for evaluating longer-term emissions of

Basket of measures	**Scenario measures studied**	**Dray et al. [11, 12]**	**Owen et al. [13]**	**Hassan et al. [14, 15]**	**Schilling et al. [16]**	**Ogunsina et al. [17]**	**EASA [18]**	**Winchester et al. [19]**	**ICAO [20]**	**Randt et al. [21]**
Technology, additional technologies and biofuels	Next-generation aircraft with improved technology‡	X	X	X	X	X	X	—	X	X
	Retrofits	X	—	—	—	—	—	—	—	—
	Performance Improvement Package (PIP) of 1% per year for all aircraft	X	X	X	—	—	X	X	—	—
	Alternative fuels	X	X	—	—	—	—	—	X	—
Infrastructure measure	ATM improvements	X	X	—	—	—	X	X	X	—
Operations measure	Early aircraft retirement	—	—	X	—	—	X	—	—	—
	Increased Load factor/ reduced frequency	X	—	—	—	—	X	—	—	—
	Others*	X	—	—	—	—	—	—	—	—

Basket of measures	Scenario measures studied	Dray et al. [11, 12]	Owen et al. [13]	Hassan et al. [14, 15]	Schilling et al. [16]	Ogunsina et al. [17]	EASA [18]	Winchester et al. [19]	ICAO [20]	Randt et al. [21]
Regulatory and economic measures	Emission trading	X	—	X	—	—	X	X	—	—
	Set emission limit	—	X	—	—	—	X	X	—	—
	Fuel tax, route and airport charge	X	—	—	—	—	X	—	—	—

‡*Electric, hybrid, distributed propulsion, blended-wing body, optimised counterrotating propeller, advanced turboprop, etc.*
↗*Surface congestion management, single-engine taxi, optimised departures, reduced cruise inefficiency, optimised approach, reduced fuel reserves, reduced tankering, increased engine maintenance, increased aerodynamic maintenance, engine wash, increased turboprop use.*

Table 1.
Emission mitigation measures studied in fleet models reviewed [22].

aircraft at the global fleet level. FSDM initially used aircraft-specific fuel consumption as the performance criterion for aircraft addition to the fleet, while aircraft were retired from the fleet using functions dependent on aircraft calendar age. This aircraft retirement method was also used in some studies previously mentioned in Section 2, without inclusion of the objective functions mentioned in Section 2.

FSDM was later adapted by Oguntona et al. [24] who evaluated aircraft direct operating cost (DOC) performance in analysing the development of the global passenger aircraft fleet under a specified forecast fuel price development. The model adaptation excluded the aircraft economic retirement consideration. However, in this study, the DOC calculation method is revised while using this approach also for aircraft economic retirement considerations.

3.1 Air transportation system simplification approach

The network simplification approach used by Oguntona et al. [24] was retained, in which average distances (hereafter referred to as route groups) between and within regions were adopted.

Furthermore, the method of classifying the aircraft fleet available in year 2008 that was adopted by Randt [23] was retained while excluding aircraft types that are unlikely to be produced as well as freighter aircraft. Two generations of aircraft types were used in the model. The initial fleet aircraft clusters referred to aircraft types available in year 2008 (see **Table 2**), while next-generation aircraft (next-gen aircraft) referred to aircraft types produced afterwards (see **Table 3**). Cost improvements in the FSDM aircraft are shown in the Appendix (see **Table A-1**).

3.2 Integrated model overview

The FSDM was built in an integrated modelling environment [23]. It receives mission fuel burn values from a global fleet mission calculator (GFMC) that is based on the Base of Aircraft Data (BADA) tool [23]. The fuel performance data is then used alongside other input data in the Aircraft Lifetime Cost (ALiTiCo) module. ALiTiCo produces lifetime direct operating cost (DOC) estimates for the different entry into service (EIS) years of the FSDM aircraft. Input for ALiTiCo includes aircraft and engine characteristics as well as other parameters that vary over time and are unique to the considered FSDM flight distances. These are shown in **Table 4**.

Cluster name	Cluster acronym	Representative aircraft type
Long-range combi	LRC	Boeing MD 11
Long range heavy	LRH	Boeing 747-400
Mid-range freighter	MRF	Boeing 767-300F
Jet commuter	JC	Embraer 190
Long-range freighter	LRF	Boeing 747-400F
Turboprop commuter	TP	ATR 72-500
Mid range	MR	Boeing 767-300
Long range	LR	Boeing 777-200
Narrow body	NB	Airbus A320-200

Table 2.
Representative aircraft of the initial fleet aircraft clusters using FSDM [21].

Initial fleet aircraft	Representative next-generation aircraft type	Next-gen aircraft cluster name (acronym)	EIS year
MR	Boeing 787-8	Next-gen mid range (NGMR)	2011
LRH	Boeing 747-800	Next-gen long range heavy (NGLRH)	2012
LR	Airbus A350XWB	Next-gen long range (NGLR)	2015
NB	Airbus A320neo	Next-gen narrow body (NGNB)	2016
JC	Bombardier CS100/ Embraer E190-E2	Next-gen jet commuter (NGJC)	2016

Table 3.
Next-generation aircraft EIS [21].

	Time-dependent input	Time-independent input	
Route-independent input			
Aircraft type dependent	• EIS year • Normalised improvements in fuel costs • Normalised improvements in non-fuel COC • Normalised improvements in ADOC	• Limit of validity • MTOW • Range at LRC • Mach number • Cabin length • Cabin height • Cabin width at floor • Cabin width maximum • Take-off field length • Operating weight empty • Approach noise level • Flyover noise level • Engine dry weight	• Number of engines • Engine SFC • Number of shafts • Engine take-off power • Number of propeller blades • Propeller diameter • BPR • OPR • Number of compressor stages • Maximum thrust • Emitted pollutant NOx per LTO cycle
Environmental, macro-economic and flight-related	• Fuel price • Inflation adjustment factors • Seat load factor • Freight load factor	• Aircraft price scenario: minimum, mean or maximum • Exchange rate • Interest and insurance rates per year • Labour rate • Escalation factors • Navigation unit rate	• Threshold approach noise • Threshold flyover noise • ETS allowance per ton CO_2 • Calculation age limit • Depreciation period
Route-dependent input			
		• Flight distance	• Flight type

Time-dependent input	Time-independent input
• Annual frequency	• Installed seats
	• Belly-freight capacity
	• Block fuel
	• Block time

Source: Own depiction.

Table 4.
Input to ALiTiCo.

An overview of the interlinked submodules of the updated integrated modelling approach used in this study is shown in **Figure 1**.

The integrated model depends on input data extracted from various data sources and processed in calculators to give various outputs. The final output of interest is obtained from the FSDM. The data flow in and out of the calculators is shown in **Figure 2**.

The main steps of the integrated model are shown in **Figure 3**.

The main steps of fleet renewal/development FSDM performs every calculation year are:

i. Retirement of aircraft that have reached their design life limit, evaluated in terms of their structural retirement age [years] depending on their assumed annual utilisation rates.

ii. Evaluation of aircraft before addition to fleet to fill route capacity gap.

iii. Comparison of existing aircraft with new available aircraft in terms of operating cost performance, retiring those having a cost disadvantage and replacing them with more cost-efficient ones. This is also referred to as economic retirement of aircraft.

3.3 Cost modelling approach

Direct operating costs (DOC) were estimated to comprise of cost of ownership (COO), cash operating costs (COC) and additional direct operating costs (ADOC) according to the approach of Ploetner et al. [25].

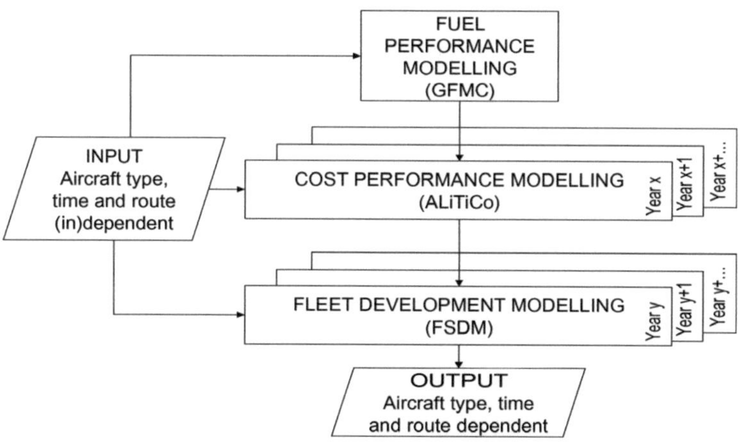

Figure 1.
Interlinked submodules of integrated model environment.

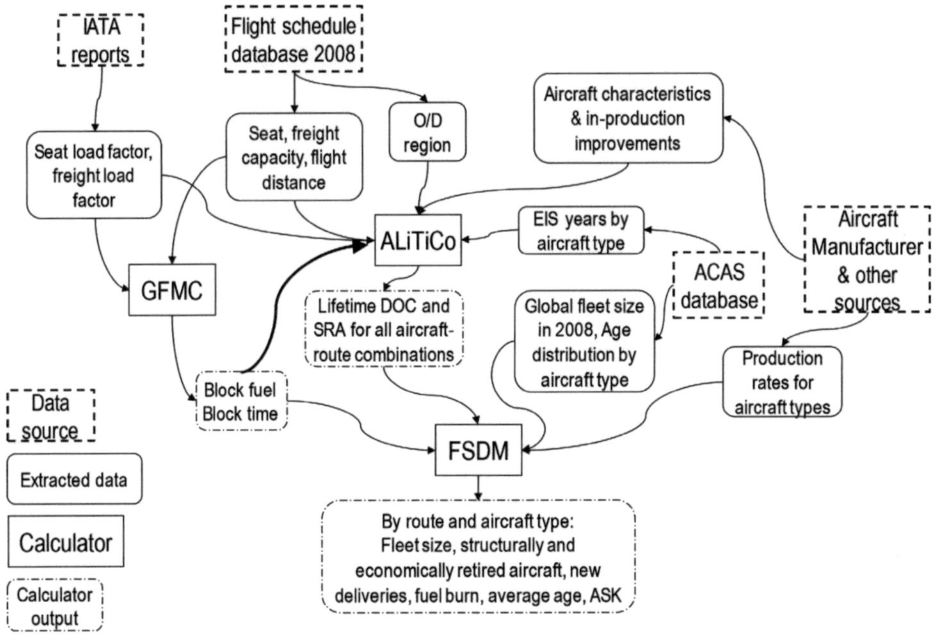

Figure 2.
Data flow in and out of GFMC, ALiTiCo and FSDM.

3.3.1 Cost of ownership

The methodology of Ploetner et al. [26] is used in this work. This approach calculates aircraft market price using aircraft parameters of range, Mach number, number of passengers, cabin volume and take-off field length based on data from year 2003 to 2008 and adjusted to year 2008 US dollars (USD). Using the relevant inflation factor [27], the costs were converted to year 2016 USD.

Given that the COO comprises of depreciation, interest and insurance, the COO development of an aircraft over time is dependent on the depreciation model chosen. In this study an exponential function model is used based on the approach of Wesseler [28]. A summary of depreciation periods of aircraft according to

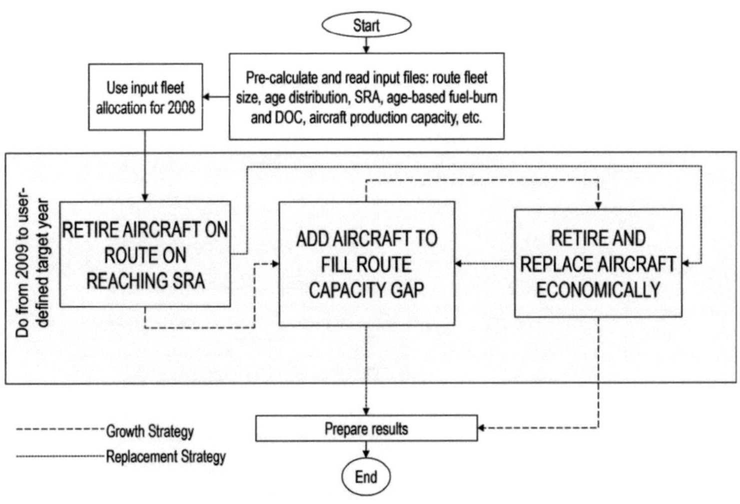

Figure 3.
Main steps in updated integrated model environment.

Literature source	Narrow-body aircraft	Wide-body aircraft
Association of European Airlines	14 years	16 years
Doganis	8–10 years	14–16 years
IATA	Average of 20 years	Average of 20 years

Table 5.
Summary of depreciation periods according to literature findings [29–32].

literature findings is shown in **Table 5**. In summary, depreciation periods of 8–20 and 14–20 years could be used for single-aisle and twin-aisle aircraft, respectively.

In ALiTiCo, aircraft delivery price is assumed to be constant over the simulation period. This is not the case in reality since aircraft prices are influenced by many factors including inflation, developments in price of materials, demand for aircraft, current market value and the strength of the dollar, among other factors [33]. However, the assumption simplifies the complexity of incorporating such effects. Furthermore, depreciation and interest period are assumed to be constant.

3.3.2 Cash operating cost

COC is composed of crew charges, fuel costs, direct maintenance costs (DMC), navigation charges, airport fees and ground handling charges.

3.3.2.1 Crew charges

Crew charges are based on a correlation relationship of crew salaries in 2008, supplied by EUROCONTROL [34], to maximum take-off weight (MTOW) and number of passengers on a flight. According to Wesseler [28], flight crew costs per block hour could be expressed as a function of MTOW, whereas cabin crew costs per block hour could be expressed as a function of the number of passengers. He assumed a seat density and combination of flight and senior flight attendants of a typical full-service carrier. Likewise, these costs were converted to 2016-year dollars, using the relevant inflation factors.

3.3.2.2 Fuel costs

Fuel costs are computed from fuel consumption and the respective yearly fuel prices, adjusting to 2016-year dollars. Fuel burn per trip is still modelled using the Global Fleet Mission Calculator (GFMC) based on the BADA 3 tool of EUROCONTROL also described by Randt [23]. The tool was already validated by Ittel [35]. However, given that fuel costs cover a major share of aircraft DOC, verification of fuel consumption estimates was done for the initial fleet and next-generation aircraft types considered in the model. Fuel burn on routes was not modelled to increase because of payload increase. However, conservatively higher passenger and freight payload factors of 86% and 53%, respectively, than in 2008 were assumed throughout the simulation period to ensure conformity of model results to anticipated future growth in these load factors, as verified by Randt [23].

In ALiTiCo, similar to the approach of Moolchandani et al. [36], engine overhaul or replacement is not done. Though fuel burn deterioration is mainly engine-driven, and thus does not have a linear characteristic throughout an aircraft's life, in ALiTiCo, a linear deterioration rate of 0.1% per year is assumed for simplification purposes. Considering that a deterioration of 3.5–4% is possible all through the aircraft life [37, 38], this assumption of 0.1% per year is justified.

3.3.2.3 Navigation charges

Navigation charges are based on the EUROCONTROL model using the average unit rate weighted by the number of landings in all European countries in 2008 [28]. The charges are computed in year 2016 US dollars (USD).

3.3.2.4 Airport charges

Airport charges and ground handling charges are based on the methodology of Ploetner et al. [39]. Airport charges are composed of landing charges, passenger charges, navigation aid charges, lighting charges, terminal charges and service charges. The method was based also on data from 2008. Like other cost components, the charges are computed in year 2016 US dollars.

In the fleet model, flights could be either within a region or between two regions. For flights belonging to the latter category, the average value of airport charges within both origin and destination regions is used.

3.3.2.5 Direct maintenance costs

Using available data from Aircraft Commerce [40] for the initial fleet aircraft, the method recommended by the Association of European Airlines (AEA method) [31, 32] is adopted because it uses aircraft parameters such as aircraft operating weight empty, engine bypass ratio, etc. Other parameters such as aircraft price are obtained from the ownership cost model already explained. Furthermore, for the engine price [year 1989 USD], the approach by Jenkinson et al. [41] is used, which calculates engine price in year 1995 British pounds based on specific fuel consumption [lb/lbf/h] and cruise thrust [Ma]. The engine bare price [year 1989 USD] is then obtained after the price in year 1995 British pounds is first converted to year 1995 USD and then to year 1989 USD.

The AEA method assumed mature levels of cost, i.e. after 5–7 years of operation. Using ageing function from Strohrmann [42], based on Dixon [43], DMC values for other years of the aircraft lifetime are determined. Furthermore, input labour rate value given by the AEA in 1989 is used and converted to 2016 USD. Due to lack of data, this is assumed to be constant over time and independent of route although DMC labour rate varies over time and with world region [44]. A limitation of the AEA method is that it does not hold for engines with thrust above 30 metric tonnes. Furthermore, since the method was developed to give comparable results to aircraft operated by airlines in 1989, the method cannot be directly used for next-generation aircraft considered in this work. Therefore, improvement factors are used which correlate nonfuel COC of initial fleet aircraft to next-generation ones.

Since the AEA method for computing aircraft DMC is evaluated in year 1989 USD, an inflation factor is used to adjust the costs to year 2016 USD. Direct maintenance costs per flight cycle of representative aircraft of the initial fleet, determined using AEA method, were compared with corresponding cost values published by Aircraft Commerce (ACC). This is shown in **Figure 4**.

Cost values published by ACC can be taken as representative of the industry since they are obtained from maintenance providers.[1] The difference between AEA and ACC values increased with increasing MTOW. A higher difference can be expected for aircraft with first flights made after the AEA publication. From **Figure 4**, the AEA method for DMC computation produced aircraft DMC values at most 22% higher than those of ACC. Compared to cost levels given by IATA's MCTF [45], the

[1] Correspondence on 13 February 2018 with Aircraft Commerce.

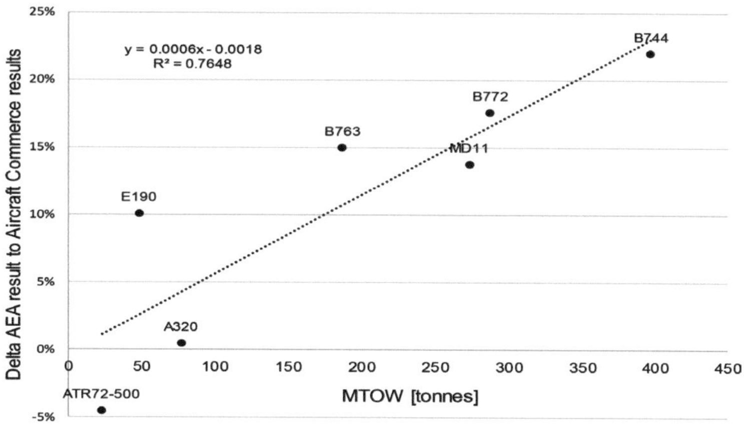

Figure 4.
DMC [$/FH] of initial fleet aircraft, comparing AEA and ACC results.

costs computed using AEA method are at most up to 40% higher. Therefore, for all initial fleet aircraft types used in FSDM, including aircraft with engine thrust above 30 metric tonnes like the B777-200, by applying a correction factor defined by the linear regression function in **Figure 4**, the cost results of the AEA method are adjusted to cost levels, resulting from Aircraft Commerce computation.

3.3.3 Additional direct operating costs

Additional direct operating costs refer to environmental airport noise and NO_x charges, as well the emission trading scheme (ETS) charges [26]. The charges were computed based on functions from Ploetner et al. [39]. The noise charges are based on defined levels of aircraft noise values for arrival as well as sideline and flyover as given by the ICAO [46]. Maximum approach and sideline and flyover noise levels of 88 and 83 EPNdB, respectively, were used. Also, a constant charge of 10 Euros per tonne of CO_2 was implemented based on Schmidt et al. [47].

4. Analysis

4.1 Longer-term fleet development strategies

If the order of implementing the last two of the three main steps of fleet renewal presented in sub-section B of the previous section is modified, two fleet renewal strategies can be defined on an FSDM route. These are shown in **Figure 5**.

The *Growth Strategy* prioritises aircraft allocation for serving demand growth and replacing aircraft retired at the end of their design lives, before replacing those that are retired because of their operating cost disadvantage. This strategy is assumed a status quo strategy used in the airline industry. This is because aircraft manufacturers claim that more than half of aircraft deliveries forecast for the next two decades are to accommodate growth in air travel demand (ATR, 65%; Embraer, 63% and 56%[2]; Boeing, 57%; Airbus, 63%) rather than replace existing aircraft (ATR, 35%; Embraer, 37% and 44%; Boeing, 43%; Airbus, 37%).[3] This is also

[2] For 70–130 seat jet segment and turboprop segment, respectively.

[3] Values derived from ATR's turboprop market forecast 2016–2035, Embraer's market outlook 2017, Boeing current market outlook 2017–2036 and Airbus Global Market Forecast 2017–2036.

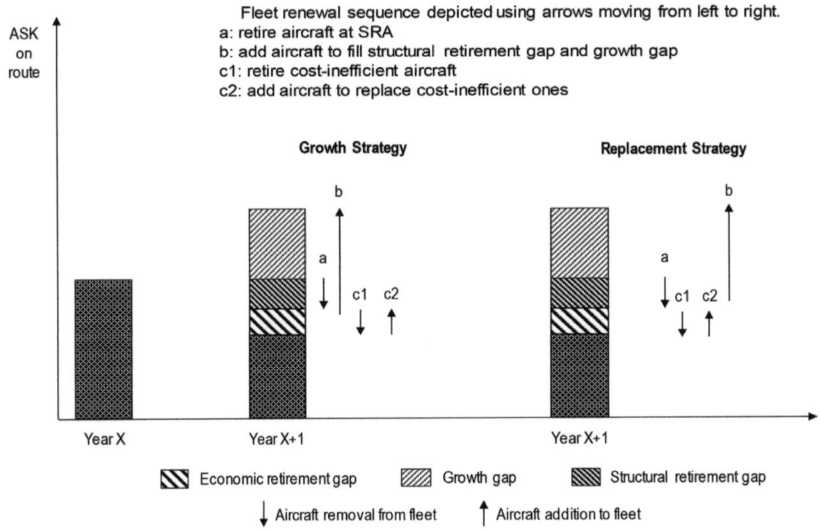

Figure 5.
Strategies for fleet renewal on a route.

assumed because airlines tend to delay replacement of their older aircraft when jet fuel prices are low [48], meaning that a complete economic retirement of aircraft has not been performed industry-wide.

Conversely, a more radical strategy, the *Replacement Strategy*, allocates global aircraft production capacity first for replacing aircraft that are retired based on their operating cost disadvantage and those retired on reaching their design life limit, before serving growth in air travel demand.

4.2 Calibration using growth strategy

FSDM was calibrated to generate fleet composition results similar to Boeing's forecast for 2036 while reflecting verifying the historical development for the jet aircraft fleet.

4.2.1 Assumed input

Passenger and freight load factors from 2008 to 2016 are taken from IATA reports [49], without differentiating between route groups. While passenger load factor increased from 76% in 2008 to 80.3% in 2016, freight load factor reduced from 46% in 2008 to 43% in 2016. In 2017, freight load factor is assumed to be slightly higher due to the entry of LCCs into the cargo business and other reasons given by JADC [50]. After 2017, freight load factor is assumed to be stable at 47.7%. However, the development in freight traffic is beyond the scope of this work. In addition, JADC [50] forecasts that passenger load factor is set to increase from 80.3% in 2016 to 83.3% in 2036.

Fuel prices (with units in year 2016 US dollars) were derived for years 1968 until 2016 using US GDP deflator values [27] and US Gulf Coast Kerosene-Type Jet Fuel Spot Price [51]. It was assumed that jet fuel prices were constant until year 1990 since there was no major difference between the average US Kerosene-Type Jet Fuel Wholesale and Resale Price by Refiners between 1978 and 1990 [52].

For years 2016–2036, low- and high-fuel price scenarios by Boeing are used as shown in **Figure 6**. The scenario of fuel price that was used in the Boeing CMO was not stated. The low-fuel price forecast assumed that fuel price in 2030 would

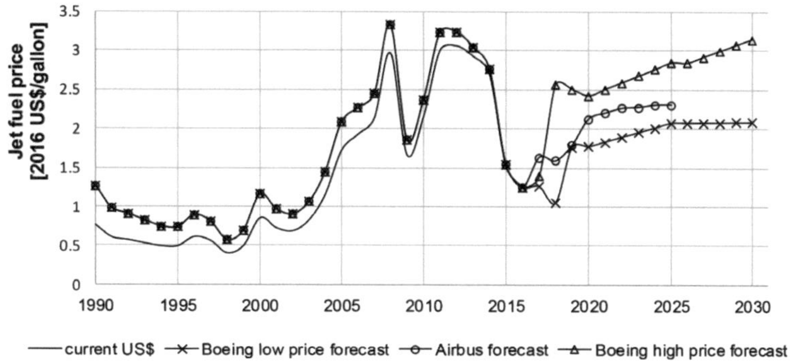

Figure 6.
Historical and forecast fuel price scenarios by Airbus and Boeing [27].

be similar to 2005 price levels. On the other hand, Boeing's high-fuel price scenario assumes that fuel price in 2018 will rise close to 2008 price level while further rising beyond 2012 price level in 2030. Same RPK growth factors on route groups were assumed for both fuel price scenarios as given in the CMO report. Fuel prices after 2030 are assumed to be stable at 2.08 and 3.1452016 US dollars per gallon in the low and high-fuel price forecasts, respectively.

These fuel price scenarios differ from Airbus fuel price forecast which assumes medium fuel price close to 2010 price levels in 2025 [53] (see **Figure 6**).

Maximum aircraft upgauge was assumed to be 20%. In addition, the utilisation of an aircraft type is modelled to vary between route groups, with a possibility of increasing annually. According to Boeing [54], passenger airplane utilisation increased between 2008 and 2015. A study of the growth in airplane utilisation at aircraft cluster level between year 2008 and 2014 revealed that the growth occurred mostly for turboprop commuter, jet commuter, narrow-body and mid-range aircraft cluster between years 2008 and 2014, whereas long-range aircraft utilisation increased between 2012 and 2014 [55]. For a given year, an increase in airplane utilisation results in lower unit costs and trip costs because fixed ownership costs are spread over an increased number of trips [56]. Considering that portions of flight crew and cabin crew costs as well as maintenance costs are possible components of fixed costs [9], it is assumed that higher airplane utilisation results in lower direct operating costs. Therefore, the unit DOC of a particular aircraft type with the same payload varies with different levels of utilisation on different route groups.

A fleet forecast also uses an assumption on development of aircraft productivity, which, according to Evans and Johnson [57], is influenced by load factor, average block speed, annual utilisation and number of seats per aircraft. Because the updated FSDM is not capable of modelling dynamically changing average block speed or number of seats per aircraft every year, aircraft productivity growth is modelled as growth in annual flight frequencies.

In their forecast, Boeing assumed older aircraft would have lower utilisation compared to newer aircraft [58]. Although an increase in passenger load factor is expected as explained above, additional annual ASK productivity growth of aircraft is modelled for next-generation aircraft as 0.9% as used by Evans and Johnson [57], with the exception of next-generation regional aircraft assumed to have a higher annual growth rate of 1.3%. For initial fleet aircraft, a lower annual growth rate of 0.35% is assumed according to Boeing's assumptions. The 0.35% growth rate is adopted from the assumption of Forsberg [59]. These values are considered conservative when considering the compound annual growth rates of initial fleet aircraft productivity between 2008 and 2014 as evaluated by Bellhäuser [55].

Lastly, after 2022, production capacity of all aircraft types is assumed to grow at an annual rate of 4.7%, same as Boeing's projected worldwide growth rate for air passenger traffic. This growth rate is arguably reasonable because over the period from 2008 to 2016, total aircraft production capacity has also grown at an average of 4.7% per year.

4.2.2 Model calibration objective

Boeing assumed in their forecast that some trends would continue. For example, they assumed that new markets that had previously been either unreachable or unprofitable, especially those that can be served by small wide-body aircraft, would open up [58]. Although the opening up of new markets is not modelled in FSDM, the effect of liberalisation, in terms of increased air traffic, is considered. These assumptions led to a forecast that the share of wide-body aircraft would increase from 19% in 2016 to 21% in 2036. As shown in **Figure 7**, this growth is driven by the growth in small twin-aisle aircraft.

This calibration work, therefore, has an objective goal of a higher preference for wide-body aircraft over narrow-body aircraft in 2050. Boeing categorised aircraft types into three groups. However, the method used in doing this was not explained. **Table 6** shows how FSDM aircraft clusters compare to the classifications.

Boeing also categorised the B777X, A350–1000 and B787–10 as M/L-TA aircraft which would be already in operation in year 2036. In the fleet model, however, these future aircraft types are not modelled as unique representative aircraft. This is mainly because the fuel burn performance of these aircraft types cannot be determined using the BADA version used in this work. Besides this, there is uncertainty about the future production capacities of these future aircraft types. Production capacities were assumed for the B777X and included in that of the NGLR, whereas, since the other two aircraft types are related to existing FSDM representative aircraft, special production capacities are not included. As a result, for calibration purposes, the production capacity share of the B777X in the NGLR is deducted from the total delivered aircraft in this aircraft cluster and its corresponding aircraft category (i.e. the S-TA) and added to the number of aircraft belonging to the category M/L-TA.

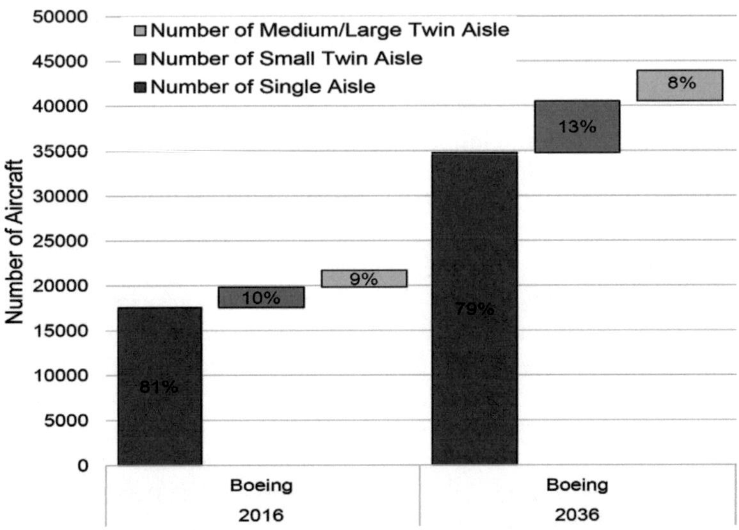

Figure 7.
Fleet size and composition according to Boeing in 2016 and 2036 [58].

Boeing category	FSDM aircraft cluster
Single aisle (SA)	JC, NB, NGJC, NGNB
Small twin aisle (S-TA)	MR, NGMR, NGLR
Medium/large twin aisle (M/L-TA)	LRC, LRH, LR NGLRH

Table 6.
Comparison of Boeing CMO to FSDM aircraft classification.

Fleet composition in year 2036 is dependent primarily on the choice of aircraft during introduction to fill the capacity gap. Apart from cost improvements modelled in the aircraft, aircraft preference depends on fuel price (FP), depreciation period (DP) and planning horizon (PH) assumed during aircraft evaluation. To have a simplified approach in calibration, these variables are applied without differentiating between single- and twin-aisle aircraft. Upper and lower boundaries for the variables are shown in **Table 7**.

Because of the high number of combinations possible if intermediate values of these variables are observed, simplifications are made in the calibration process by using only combinations involving the boundary values. Moreover, further assumptions were made in terms of cost improvements due to the increase in aircraft utilisation, leading to a preference for S-TA and M/L-TA. These are shown in the Appendix (see **Table A-2**). For each fuel burn scenario, four combinations of DP and PH are used for calibration.

4.2.3 Calibration results

The calibration results are shown in **Figure 8** for years 2008, 2016 and 2036. Because the long-term development is of interest, yearly changes in the results are not shown.

In 2008, FSDM's total fleet size, which is taken from the ACAS database, was 3% less than that of Boeing because the latter considered more aircraft types (e.g. twin-aisle aircraft like Ilyushin IL-86 and Lockheed L-1011 and single-aisle aircraft like Sukhoi Superjet 100,Yakovlev Yak-42, Mitsubishi MRJ, Dornier 328JET, Fokker 70, F28 and BAe 146). As a result, FSDM's SA fleet size was 10% less than that of Boeing. When weighted by their 75% share in the total aircraft fleet size, a difference of 7% results. Furthermore, for the S-TA and M/L-TA aircraft categories, respectively, with approximately 12% share each, percentage differences in fleet size estimates by Boeing and FSDM of 60 and 4% were estimated. When weighted by fleet size, differences of 7 and 0% result for the S-TA and M/L-TA, respectively. Therefore, a maximum difference of 7% is estimated between the fleet sizes in each category, when weighted by fleet share, based on Boeing's data and those of the FSDM simulation year.

In 2016, some of these aircraft types, especially single aisles, not considered in FSDM initial fleet were in limited service. Therefore, the share of single-aisle aircraft

Variable	Low boundary	Upper boundary
Fuel price scenario	Boeing low fuel price	Boeing high fuel price
Depreciation period	14 years	20 years
Planning horizon	7 years	15 years

Table 7.
Upper and lower boundary values of calibrated variables.

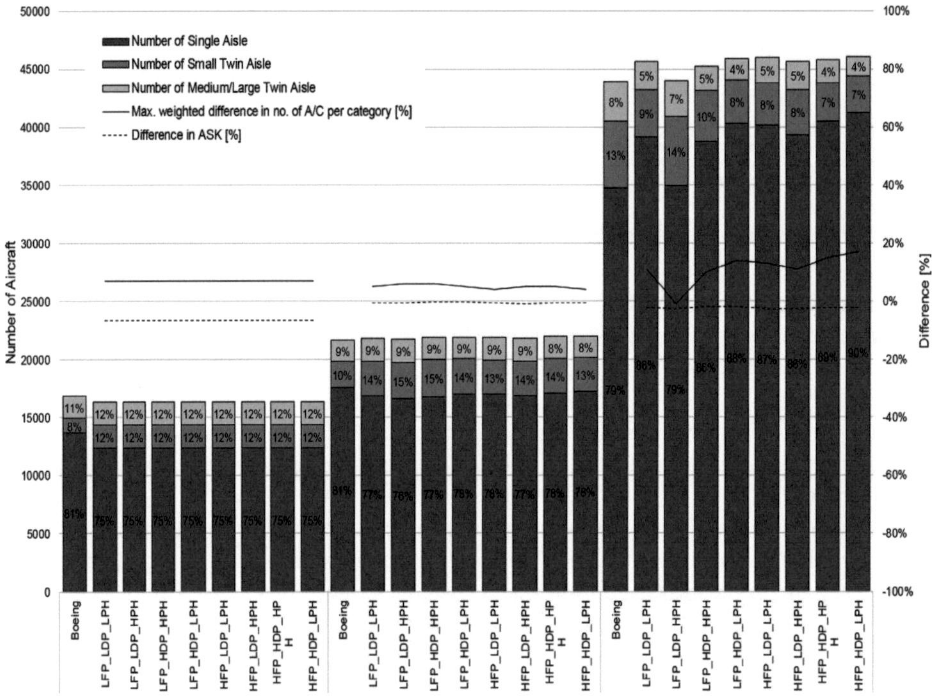

Figure 8.
Calibration results.

between 2008 and 2016 is expected to have increased slightly over the period. As a result, in 2016, FSDM produced a slightly higher share of single-aisle aircraft compared to 2008, although the share of single-aisle aircraft did not change in Boeing's data from 2008 to 2016. In 2016, the difference between the numbers of aircraft in each category based on Boeing's data and those of the FSDM in 2016 ranges between −6 and 43%. However, when weighted by fleet share, there was a 6% maximum difference in fleet size estimates by FSDM and Boeing for each aircraft category.

In 2036, the difference in fleet size estimates by FSDM and Boeing for each aircraft category ranged from −51 to 19% depending on the combination of calibration variables used. However, when weighted by fleet share, there was a maximum difference of 17% in fleet size estimates by FSDM and Boeing for each aircraft category. The calibration results in terms of ASK show a good comparison to Boeing's forecast. In years 2008, 2016 and 2036, maximum differences in ASK of −7, −1 and −3%, respectively, were attained.

For both fuel price forecasts used, of the four combinations of DP and PH possible, the combination of low depreciation period (LDP) and high planning horizon (HPH) gives results closer to Boeing's forecast. Therefore, this combination is used in the remaining steps of this study. The most comparable result to the jet fleet composition forecast by Boeing is obtained using the low-fuel price (LFP) scenario in the LDP and HPH combination. In other words, this combination has the lowest maximum difference between the numbers of aircraft in each category based on Boeing's data and those resulting from the FSDM in 2036.

Furthermore, from **Figure 8**, it can be seen that an increase in jet fuel price from Boeing's low- to high-price scenarios leads to a "slightly" different fleet composition in 2036. This primarily results from a change in the ranking and introduction of cost-efficient aircraft on the route groups, leading to an increase in the number and share of narrow-body aircraft. Because, unlike wide-body aircraft, narrow-body aircraft are less sensitive to fuel price, a higher jet fuel price has less impact in increasing their unit DOC. As a result, narrow-body aircraft are more competitive than their wide-body counterparts are, especially when compared on the design range of the former. This result is in agreement with the claim by Rutherford [60] that aircraft with four engines like the B747 and A380 were less fuel-efficient than fuel-efficient twinjets like the A350–900 and B787–9 even on trans-Pacific routes for which the former are designed. Therefore, in a high-fuel price scenario like that of Boeing, fleet phase-in decisions will favour less of NGLRH. Furthermore, given that fuel prices change over time, the comparative cost performance of the aircraft differs over time.

4.3 Verification using growth strategy

4.3.1 Verification of historical fleet fuel burn and fuel efficiency

Based on calculation results by Wasiuk et al. [61], IATA [49, 62] and Dray et al. [63], estimates of past passenger aircraft fuel burn in million tonnes are compared with results from FSDM. FSDM estimates are in average 5% above the estimates of Dray et al. and 7% below IATA's estimate for year 2016. However, FSDM's estimate of fleet fuel burn in 2016 is 1% below that of the IATA. Using the approach explained by Dray et al. [63], IATA's values used here are reduced by 9.6% because, unlike the IATA reports, freighter aircraft are not included in this work. Another 5% was deducted to account for unscheduled flights that were included in IATA reports. For the fuel efficiency results, IATA's fuel burn data was combined with ICAO's ASK data. In 2016, FSDM exactly reproduces fuel efficiency data by the IATA and ICAO. Fuel burn performance of the global passenger aircraft from year 2008 to 2016 is shown in **Figure 9**.

4.3.2 Verification of historical fleet unit cost and average age

Fleet unit cost development is dependent on development in fuel unit cost [64]. The developments in cost per ASK (CASK), fuel price and average aircraft age are shown in **Figure 10**.

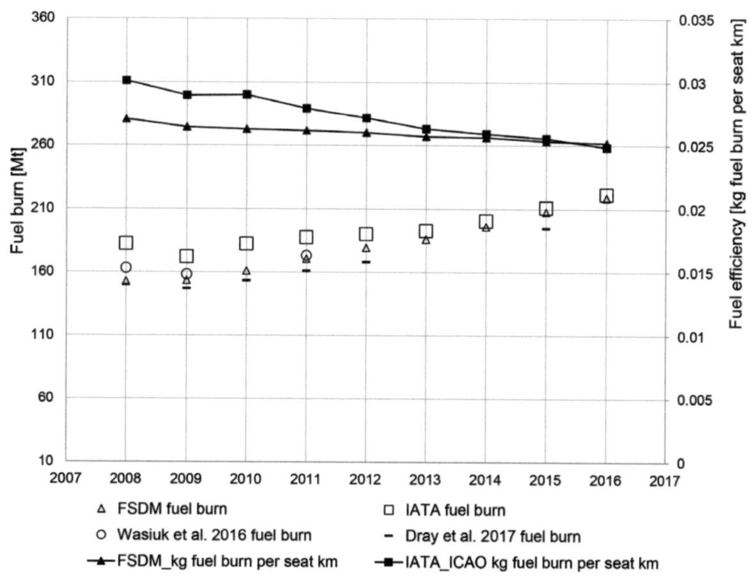

Figure 9.
Passenger aircraft fuel burn and fuel efficiency 2008-2016.

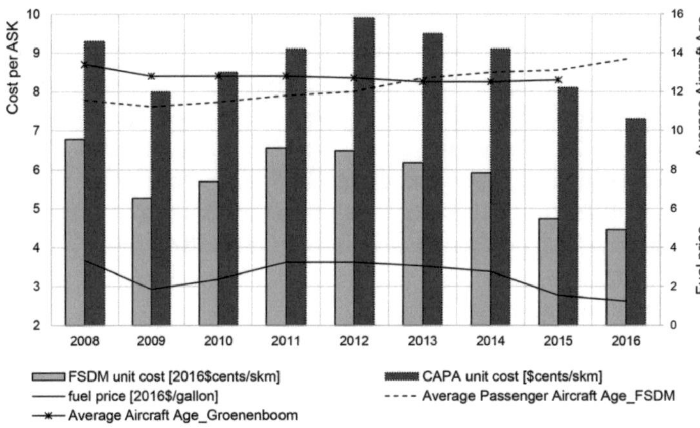

Figure 10.
Fuel price, unit cost, and average age of global pax fleet: 2008-2016. Source: Own calculations [65].

Data on aircraft unit cost used for verification was obtained from Centre for Asia Pacific Aviation [64]. A comparable unit cost drop between 2014 and 2016 can be observed for both FSDM results (25%) and CAPA data (20%). Likewise, the trend of the cost development throughout the period is comparable for both FSDM and CAPA, although absolute values are not equal. Although Groenenboom [65] recorded that the average age of passenger aircraft slightly decreased between 2010 and 2015, it does not precisely give the age for passenger aircraft. Average age of the passenger aircraft fleet depends on the rate of aircraft additions to the fleet, compared to retirements from the fleet. In addition, the average fleet age depends on jet fuel price. Lower fuel prices encourage airlines to keep older aircraft longer in service, especially when travel demand is strong [66, 67], thereby increasing the average fleet age. Therefore, a slight increase in the average age of the fleet accompanies a decrease in the price of fuel from year 2012 to 2016.

4.3.3 Verification of forecast fleet fuel burn and air passenger traffic

Next, the reliability of the model in estimating future emissions and air passenger traffic of the global passenger aircraft fleet is verified.

For forecasts until 2050, passenger load factor is assumed steady at 2036 levels. Dray et al. [63] updated AIM to AIM2015 and used the UK Department of Energy and Climate Change (DECC) historical and forecast oil price levels [68]. In this verification study, the DECC medium oil price forecast was used. A review of the historical prices [year 2016 USD per gallon] between 1990 and 2015 shows that jet fuel prices were approximately 21% above DECC oil prices. The fuel price development according to the DECC has a price level in 2036 and beyond which is even higher than Boeing's high-fuel price forecast.

Furthermore, from year 2015, RPK growth rates of 3.8% per year were used in this verification process according to the SSP2 baseline scenario of Dray et al. [63]. In the SSP2 baseline scenario, zero carbon prices were assumed, so that ETS costs were set to zero. The assumptions in aircraft utilisation, load factor and technology improvements used for arriving at Boeing's future fleet composition are retained. As a result, the basic giant-leap technological improvements assumed were similar. Incremental improvements were excluded since they did not assume incremental technological improvements. **Figure 11** shows the jet fuel price development of the SSP2 baseline scenario.

Figure 12 shows estimates of fuel burn and air traffic in 2050 relative to 2015 from Dray et al. [63] and using FSDM. Because the long-term development is of interest, yearly changes in the results are not shown.

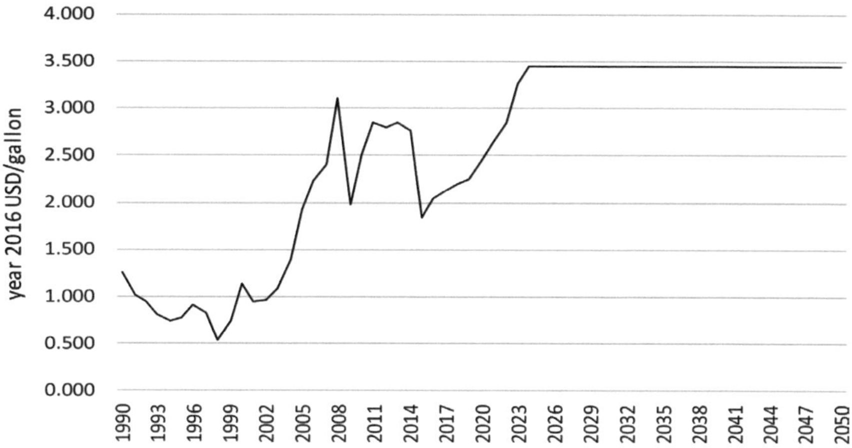

Figure 11.
Jet fuel price development in the SSP2 baseline scenario. Source: Own calculations, based on [63].

Dray et al. [63] obtained a range of results in the fleet fuel burn depending on the modelled scenario for future technology. FSDM's forecast fuel burn falls between their forecast boundaries as shown in the figure. Furthermore, there is a slight difference between the relative developments of RPK for the two models. This may be due to different input assumptions on aircraft size and utilisation used in both models as noted by Dray et al. [63]. However, since the fleet size, composition and capacity forecast ability of FSDM have been tested, this difference can be neglected.

4.4 Emission mitigation potential of the retirement strategy

4.4.1 General input for analysis

Past and forecast RPK growth factors are used as given by Boeing [58]. After 2036, the annual growth rates are assumed constant at 2036 levels. Assumptions on seat and freight load factor are the same as in the verification according to Boeing's forecast. Past fuel price until 2016 and forecast prices by Airbus until 2025 are used

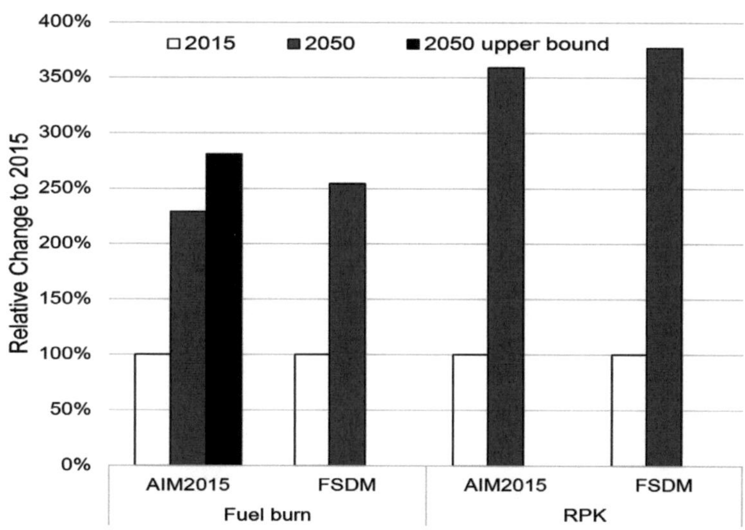

Figure 12.
Verification of forecast passenger aircraft fuel burn and traffic. Source: Own calculations, based on [63].

as shown in **Figure 6**. Fuel price after 2025 is assumed to increase annually by 0.1%, reaching 2.53 year 2016 US dollars per gallon in year 2050.

Fleet planning horizon and aircraft depreciation period are kept at 15 and 14 years, respectively. Aircraft production capacity and annual productivity are as earlier described. Furthermore, calibration input such as cost improvement assumptions is retained for all application scenarios.

5. Definition of scenarios

A baseline scenario is assumed which is named *giant-leap plus incremental improvement baseline*. This scenario assumes that incremental improvements are applied to initial fleet and next-generation aircraft. Thus, it assumes that airlines always integrate the latest available fuel and cost improvements on aircraft programmes when adding new available aircraft to the fleet. The actual state of CO_2 emissions from passenger air transport is expected to be similar to this baseline if in-service improvements on aircraft otherwise known as Performance Improvement Packages (PIPs), which are beyond the scope of this research, are considered. Incremental improvements refer all improvements shown in the Appendix (see **Table A-1**), excluding those comparing each aircraft to its previous aircraft generation.

Whereas the *Growth Strategy* has been used in the FSDM calibration and verification and is the strategy used for the baseline scenario, the *Replacement Strategy* is evaluated as a strategic measure to determine emission reduction benefits at the fleet level. Other assumptions of the baseline scenario are kept, except the fleet renewal sequence.

6. Results

Prioritising filling retirement gaps above growth gap implies that more aircraft production capacity is used for replacing economically inefficient aircraft. Compared to the *Growth Strategy*, the *Replacement Strategy* generates a higher wave of aircraft economic retirement. Between 2008 and 2050, the *Replacement Strategy* retires 7% more aircraft economically than the *Growth Strategy*. Between 2008 and 2024, the *Replacement Strategy* retires approximately 65% more aircraft economically and 44% more aircraft both economically and structurally. This can be seen in **Figure 13**.

However, in year 2024, few years after the JC, MR, LR and NB would be out of production; only a 3% improvement in CO_2 emissions is realised using the *Replacement Strategy* compared to the *Growth Strategy*.

From **Figure 14**, the two benefits of the *Replacement Strategy* until 2024 can be seen—a maximum of 2% higher share of retired aircraft in the fleet and a slightly longer year-on-year growth in fleet specific fuel consumption (SFC). However, because these improvements are minimal, the CO_2 emission improvement in the *Replacement Strategy* is also limited to about 3% in year 2024.

However, after year 2024, having attained a more cost-efficient and fuel-efficient fleet than in the *Growth Strategy*, the growth in ASK over time and the absence of more efficient aircraft result in fewer numbers of aircraft being retired by the *Replacement Strategy*. On the other hand, in the *Growth Strategy*, the fleet in year 2024 is not as efficient, thereby giving a possibility of better fleet renewal afterwards. Between 2025 and 2050, the *Growth Strategy* retires 4% more aircraft both economically and structurally than the *Replacement Strategy*. Therefore, compared to the *Growth Strategy*, the *Replacement Strategy* gives a lower EMP of 2% in 2050.

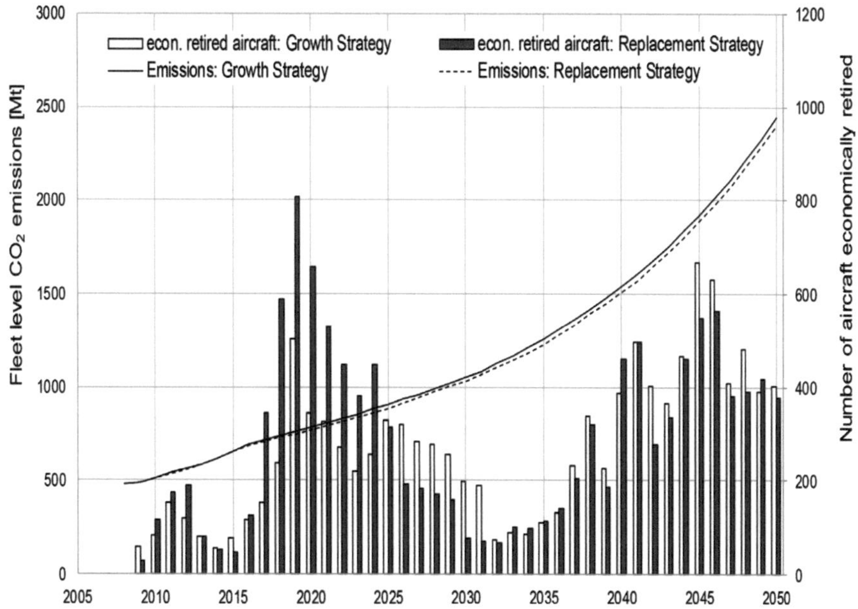

Figure 13.
Growth and replacement strategy: Number of aircraft economically retired and fleet level CO_2 emissions.

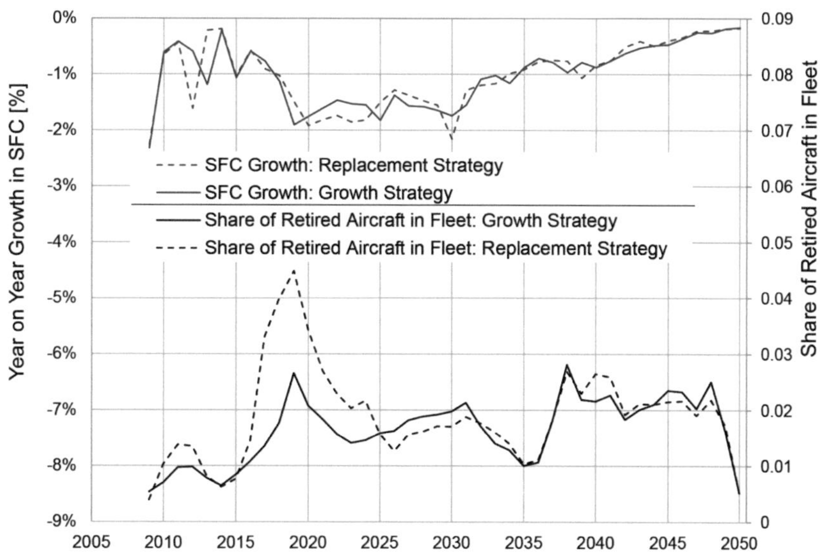

Figure 14.
Growth and replacement strategy: Share of retired aircraft in fleet and year-on-year growth in specific fuel consumption.

7. Conclusion

From the last fleet-level emission results of **Figure 13**, there is an identified short-term benefit of the *Replacement* Strategy. However, in the long term, the benefits reduce because the absence of newer more cost-efficient aircraft. Despite this diminished benefit, about 2% of the emissions (40 Mt. CO_2) could be saved at the global fleet level in year 2050 by using this fleet renewal strategy. Therefore, in order to achieve higher mitigation potentials, there is need for additional technological measures, in terms of more cost-efficient aircraft, with entry to service as from 2024.

As an outlook, having known the emission mitigation potential of the proposed *Replacement Strategy*, the cost analysis, i.e. advantage or disadvantage, of this measure should also be evaluated. Lastly, this study does not include effects of ticket price elasticity of fuel price changes.

Acknowledgements

This research was carried out using a Munich Aerospace Scholarship. The author would like to thank Professor Mirko Hornung and Dr. Kay Ploetner, as well as other members of the Munich Aerospace network, for their useful comments made during the research.

Appendix A

See Tables A-1 and A-2.

Aircraft cluster (earliest EIS year)	**Cost improvements from cluster introduction to service [%]**		
	Fuel cost	**Nonfuel COC**	**ADOC**
Long-range combi (1973)	Cruise performance improvement package: −4% [69]		
Long range heavy (1973)	Interior changes, aerodynamics and engine: −25% [70] Engine: −12% [70] PW4000 94-inch upgrade package: −1% [71]	B747–400 entry into service: −26% [72, 73]	
Jet commuter (1968)	E190 EIS, −18% [74, 75]; aerodynamic enhancements, −2% [76]	E190 improvement compared to previous aircraft in cluster, −25%; maintenance improvement, −5% [77, 78]	
Turboprop commuter (1990)	Use of ARMONIA cabin: −0.6% [79]		
Long range (1991)	PIP, −1% [80]; upgrade package, −2% [81]	−3.4% [82]	−3.4% [82]
Narrow body (1968)	A320 improvement compared to previous aircraft in cluster: −16.6% [83, 84] Other improvements including wingtip fence and sharklets: −3.5% [82]	Compared to previous aircraft in cluster: −7.9% [83] Average improvement adopted from B737–800: −2.5% [82]	Average improvement adopted from B737–800: −2.5% [82]
Next-gen mid range (2011)	Compared to previous generation, −20% [85, 86]; Trent 1000 TEN, −2% [87, 88]	Compared to previous generation: −10% [89]	
Next-gen long range heavy (2012)	Compared to previous generation: −16% [90], −3.5% [91]	Compared to previous generation: −3% [73]	

Aircraft cluster (earliest EIS year)	**Cost improvements from cluster introduction to service [%]**		
	Fuel cost	**Nonfuel COC**	**ADOC**
Next-gen long range (2015)	Compared to previous generation, −25% [92]; Trent XWB-84-enhanced performance, −1% [93]; sharklets, −1.4% [94]	Compared to previous generation: −25% [92]	
Next-gen narrow body (2015)	Compared to previous generation, −15% [91]; PW engine improvement, −2% [92]	Compared to previous generation: −14% [95]	
Next-gen commuter (2016)	Compared to previous generation: −17.3% [96]	Compared to previous generation: −10% [97]	Lower noise: −2% [98]

Table A-1.
Incremental and giant-leap cost improvements of FSDM aircraft types.

Aircraft cluster	**Cost improvements [%]**		
	Fuel cost	**Nonfuel COC**	**ADOC**
LRH	−4%	−7%	
MR	−1.8%	−3.6%	
LR	−2%	−2%	
NGMR	−14.9%	−9.8%	
NGLR	−15%	−5%	

Table A-2.
Additional aircraft cluster cost improvements assumed during calibration.

Author details

Oluwaferanmi Oguntona
Munich Aerospace e.V., Faculty of Aerospace Engineering, Taufkirchen, Germany

*Address all correspondence to: ooguntona@gmail.com

References

[1] Mowry JJ. Retirements and Inductions—How are Fleet Demographics Changing? 2018. Available from: altonaviation.com/wp-content/uploads/2018/02/2018-Aero-Engines-Miami-Alton-Fleet-Retirement-Trends-01Feb2018.pdf22.04.2019 [22 April 2019]

[2] Air Transportation Action Group, Facts & Figures. 2018. Available from: www.atag.org/facts-figures.html22.04.2019 [22 April 2019]

[3] Japan Aircraft Development Corporation. Worldwide Market Forecast: 2018-2037. 2018. Available from: www.jadc.jp/files/topics/140_ext_01_en_0.pdf22.04.2019 [22 April 2019]

[4] ICAO. FAQ 1. Why has ICAO decided to develop a global MBM scheme (GMBM) for international aviation? n.d. Available from: www.icao.int/Meetings/HLM-MBM/Pages/FAQ1.aspx16/11/2016 [16 November 2016]

[5] IATA. Fact Sheet. Climate Change and CORSIA. 2018. Available from: www.iata.org/pressroom/facts_figures/fact_sheets/Documents/fact-sheet-climate-change.pdf22.04.2019 [22 April 2019]

[6] Air Transportation Action Group. The right flightpath to reduce aviation emissions. 2010. Available from: www.atag.org/component/attachments/attachments.html?id=7204/09/2016 [04 September 2016]

[7] SGI Aviation. Aircraft Decommissioning Study: Final Report. 2018. Available from: www.sgiaviation.com/wp-content/uploads/2018/05/IATA_Aircraft_Decommissioning_Study_May-2018.pdf22.04.2019 [22 April 2019]

[8] Wensveen JG. Air Transportation: A Management Perspective. 7th ed. Farnham: Ashgate Publishing Limited; 2012

[9] Clark P. Buying the Big Jets: Fleet Planning for Airlines. London: Routledge; 2017

[10] Belobaba P, Odoni AR, Barnhart C, editors. The Global Airline Industry. Chichester, UK: Wiley; 2009

[11] Dray LM, Evans A, Schäfer A. The impact of economic emissions mitigation measures on global aircraft emissions. 2010. Available from: aimproject.aero/Documents/ATIO2010Dray.pdf15.08.2016 [15 August 2016]

[12] Dray LM, Schäfer AW, Zayat KA. The global potential for CO_2 emissions reduction from jet engine passenger aircraft. Transportation Research Record: Journal of the Transportation Research Board. 2018;**43**:036119811878736

[13] Owen B, Lee DS, Lim L. Flying into the future. Environmental Science & Technology. 2010;**44**:2255-2260

[14] Hassan M, Payan A, Pfaender H, Garcia E, Schutte J, Mavris D. Framework development for performance evaluation of the future national airspace system. 2015. Available from: www.researchgate.net/publication/280114298_Framework_Development_for_Performance_Evaluation_of_the_Future_National_Airspace_System12.09.2016 [12 September 2016]

[15] Hassan M, Pfaender H, Mavris D. Feasibility analysis of aviation CO_2 emission goals under uncertainty. In: AIAA Aviation Technology, Integration, and Operations Conference; 2017

[16] Schilling T. AIRCAT Assessment of the Impact of Radical Climate-Friendly Aviation Technologies. 2016. Available from: elib.dlr.de/100416/1/AIRCAT_Report_20160429_IB.pdf 03.03.2017 [03 March 2017]

[17] Ogunsina K, Chao H, Kolencherry N, Moolchandani K, Crossley WA, DeLaurentis DA. A model of aircraft retirement and acquisition decisions based on net present value calculations. In: AIAA AVIATION Forum; 2017

[18] EASA. Study on Aviation and Economic Modelling. 2010. Available from: easa.europa.eu/system/files/dfu/2010-SAVE-Study%20on%20AViation%20and%20Economic%20Modelling-Final%20Report.pdf10.03.2015 [10 March 2015]

[19] Winchester N, Wollersheim C, Clewlow R, Jost NC, Paltsev S, Reilly JM, et al. The impact of climate policy on US aviation. Journal of Transport Economics and Policy. 2013;**47**:1-15

[20] ICAO. ICAO Environmental Report 2016. 2016. Available from: www.icao.int/environmental-protection/Documents/ICAO%20Environmental%20Report%202016.pdf03.11.2016 [03 November 2016]

[21] Randt NP, Jessberger C, Ploetner KO. Estimating the fuel saving potential of commercial aircraft in future fleet-development scenarios. In: 15th AIAA Aviation Technology, Integration, and Operations Conference. 2015

[22] Oguntona O. Longer-term aircraft fleet modelling: Narrative review of tools and measures for mitigating carbon emissions from aircraft fleet. CEAS Aeronautical Journal. 2018. under review

[23] Randt NP. Aircraft technology assessment using fleet-level metrics [PhD dissertation]. Garching; 2016

[24] Oguntona O, Cui A, Ploetner KO, Hornung M. In: Anders, editors. Fleet Development Planning of Airlines: Incorporating the Aircraft Operating Economics Factor. 2016

[25] Ploetner KO, Schmidt M, Baranowski D, Isikveren AT, Hornung M. Operating Cost Estimation for Electric-Powered Transport Aircraft. 2013. Available from: www.academia.edu/21808291/Operating_Cost_Estimation_for_Electric-Powered_Transport_Aircraft17/05/2016 [17 May 2016]

[26] Ploetner KO, Cole M, Hornung M, Isikveren AT, Wesseler P, Essling C. Influence of aircraft parameters on aircraft market price. 2012. Available from: www.researchgate.net/publication/274705788_Influence_of_Aircraft_Parameters_on_Aircraft_Market_Price10.01.2017 [10 January 2017]

[27] Arrepim. Converter of current to real US dollars. Arrepim AG; 2018

[28] Wesseler P. Determination of aircraft life cycle cost during conceptual design phase [Master]. Garching; 2011

[29] IATA. Airline Disclosure Guide: Aircraft acquisition cost and depreciation. 2016. Available from: www.iata.org/publications/Documents/Airline-Disclosure-Guide-aircraft-acquisition.pdf23.01.2018 [23 January 2018]

[30] Doganis R. Flying Off Course: Airline Economics and Marketing. 4th ed. London: Routledge; 2009

[31] Association of European Airlines. Long Range Aircraft AEA Requirements; 1989

[32] Association of European Airlines. Short-Medium Range Aircraft AEA Requirements; 1989

[33] Kiminski-Morrow D. Airbus includes surcharge in 2008 catalogue prices. 2008. Available from: www.flightglobal.com/news/articles/airbus-includes-surcharge-in-2008-catalogue-prices-223184/19.01.2018 [19 January 2018]

[34] University of Westminster. Dynamic cost indexing: Technical discussion document 5.0. Aircraft

crewing—marginal delay costs. 2008. Available from: www.eurocontrol.int/eec/gallery/content/public/documents/projects/CARE/CARE_INO_III/DCI_TDD5-0_Aircraft_crewing-marginal_delay_costs.pdf16.01.2018 [16 January 2018]

[35] Ittel J. Developing a software tool for comprehensive flight performance and mission analyses in the context of the assessment of a novel turboprop transport aircraft concept [Master]. 2014

[36] Moolchandani K, Govindaraju P, Roy S, Crossley WA. Assessing effects of aircraft and fuel technology advancements on select aviation environmental impacts. Journal of Aircraft. 2017;**54**:857-869

[37] Kelaidis M, Aretakis N, Tsalavoutas A, Mathioudakis K. Optimal mission analysis accounting for engine aging and emissions. Journal of Engineering for Gas Turbines and Power. 2009;**131**:11201

[38] Wulf RH. Engine diagnostics program: CF6-50 engine performance deterioration. 1980. Available from: ntrs.nasa.gov/archive/nasa/casi.ntrs.nasa.gov/19810003575.pdf13.03.2018 [13 March 2018]

[39] Ploetner KO, Wesseler P, Phleps P. Identification of key aircraft and operational parameters affecting airport charges. IJAM. 2013;**2**:91

[40] Aircraft Commerce. Owner's & Operator's Guide: 777-200/−300. 2008. Available from: www.team.aero/images/aviation_data_insert/Owners_OperatorsGuide_777.pdf24/01/2017 [24 January 2017]

[41] Jenkinson LR, Simpkin P, Rhodes D. Civil Jet Aircraft Design. London: Arnold; 1999

[42] Strohrmann K. Economic modelling of aircraft entry- and out-of-service behaviour for new aircraft concepts [master thesis]. Garching; 2015

[43] Dixon M. The Maintenance Costs of Aging Aircraft: Insights from Commercial Aviation. 2006. Available from: www.rand.org/content/dam/rand/pubs/monographs/2006/RAND_MG486.pdf14.11.2017 [14 November 2017]

[44] Derber A. Where MRO Labour Rates Are Heading. 2017. Available from: www.mro-network.com/maintenance-repair-overhaul/where-mro-labor-rates-are-heading16.12.2017 [16 December 2017]

[45] IATA. Airline maintenance cost executive commentary: An exclusive benchmark analysis (FY2009 data) by IATA's Maintenance Cost Task Force. 2011. Available from: www.cofemersimir.gob.mx/expediente/12761/mir/30340/archivo/86862312.12.2017 [12 December 2017]

[46] ICAO. Noise Certification Database. 2017. Available from: noisedb.stac.aviation-civile.gouv.fr/index.html12.01.2018 [12 January 2018]

[47] Schmidt M, Gologan C, Ploetner KO. A320 Reference Costing Model 2008; 2013

[48] Flottau J. A320neo And Boeing 737 MAX Spur Market Shifts. 2017. Available from: aviationweek.com/commercial-aviation/a320neo-and-boeing-737-max-spur-market-shifts22.12.2017 [22 December 2017]

[49] IATA. Economic Performance of the Airline Industry: Industry Forecast—June 2017. 2017. Available from: www.iata.org/publications/economics/Reports/Industry-Econ-Performance/Central-forecast-mid-year-2017-tables.pdf12.03.2018 [12 March 2018]

[50] Japan Aircraft Development Corporation. Worldwide Market Forecast: 2017-2036. 2017. Available from: www.jadc.jp/files/topics/118_ext_01_en_0.pdf13.04.2018 [13 April 2018]

[51] U.S. Energy Information Administration. Petroleum and Other Liquids: U.S. Gulf Coast Kerosene-Type Jet Fuel Spot Price FOB. 2018. Available from: www.eia.gov/dnav/pet/hist/LeafHandler.ashx?n=PET&s=EER_EPJK_PF4_RGC_DPG&f=A12.03.2018 [12 March 2018]

[52] U.S. Energy Information Administration. Petroleum and other liquids: U.S. kerosene-type jet fuel wholesale/resale price by refiners. 2018. Available from: www.eia.gov/dnav/pet/hist/LeafHandler.ashx?n=PET&s=EMA_EPJK_PWG_NUS_DPG&f=A12.03.2018 [12 March 2018]

[53] Airbus. Global Market Forecast: Growing Horizons 2017/2036. 2017. 01.01.2018 [01 January 2018]

[54] Boeing Commercial Airplanes. Current Market Outlook: 2016-2035. 2016. Available from: www.boeing.com/resources/boeingdotcom/commercial/about-our-market/assets/downloads/cmo_print_2016_final_updated.pdf17.01.2018 [17 January 2018]

[55] Bellhäuser LM. Aircraft productivity analysis and fleet development planning [master thesis]. Munich; 2016

[56] Mirza M. Economic impact of airplane turn-times. In: AERO Magazine; 2008

[57] Evans R, Johnson M. Market Outlook 2006-2025. 2006. Available from: www.aerohabitat.eu/uploads/media/28-11-2007_-_Rolls_Royce__Market_outlook_2006.pdf29.05.2018 [29 May 2018]

[58] Boeing Commercial Airplanes. Current Market Outlook: 2017-2036. 2017. 03.03.2018. [03 March 2018]

[59] Forsberg D. World Fleet Forecast 2017-2036. 2017. Available from: avolon.aero/wp/wp-content/uploads/2017/10/Avolon_White-paper_Digital-27.10.17.pdf

[60] Rutherford D. Size matters for aircraft fuel efficiency. Just not in the way that you think. 2018. Available from: www.theicct.org/blog/staff/size-matters-for-aircraft-fuel-efficiency18.04.2018 [18 April 2018]

[61] Wasiuk D, Khan M, Shallcross D, Lowenberg M. A commercial aircraft fuel burn and emissions inventory for 2005-2011. Atmosphere. 2016;7:78

[62] IATA. 2017 WATS: World Air Transport Statistics. Participant Copy. 2017. Available from: www.ryanair.com/content/dam/ryanair/2017/tv-campaign/WATS-2017.pdf10.3.2017 [10 March 2017]

[63] Dray L, Krammer P, Doyme K, Wang B, Kinan AZ, Aidan OS, Schäfer A. AIM2015: Validation and initial results from an open-source aviation systems model. In: ATRS World Conference; 2017

[64] CAPA. CAPA Airline Profit Outlook. 2017. Available from: centreforaviation.com/insights/airline-leader/capa-airline-profit-outlook-34009815.03.2018 [15 March 2018]

[65] Groenenboom J. Aircraft retirements and part-out: Effective use of existing trends and opportunities in the market. 2016. Available from: www.iata.org/whatwedo/workgroups/Documents/MCC-2016-BKK/D2-1000-1030-aircraft-retirement-sgiaviation.pdf20.12.2017 [20 December 2017]

[66] Canaday H. Mid-life aircraft seen differently given fuel prices, passenger demand. 2015. Available from: www.mro-network.com/maintenance-repair-overhaul/mid-life-aircraft-seen-differently-given-fuel-prices-passenger-demand16.04.2018 [16 April 2018]

[67] Seidenman P, Spanovich D. Predicting when commercial airliners will be torn down. 2017. Available from: www.

mro-network.com/airframes/predicting-when-commercial-airliners-will-be-torn-down16.04.2018 [16 April 2018]

[68] Department of Energy & Climate Change. DECC 2015 fossil fuel price assumptions. 2015. Available from: assets.publishing.service.gov.uk/government/uploads/system/uploads/attachment_data/file/477958/2015_DECC_fossil_fuel_price_assumptions.pdf18.04.2018 [18 April 2018]

[69] National Research Council (U.S.). Assessment of Wingtip Modifications to Increase the Fuel Efficiency of Air Force Aircraft. Washington, DC: National Academies Press; 2007

[70] Davis RA. 747 modernization. In: Santini P, Staufenbiel R, editors. ICAS Proceedings 1986: 15th Congress of the International Council of the Aeronautical Sciences, London. 1986. pp. 97-107

[71] Pratt & Whitney. Pratt & Whitney Offers PW4000 94-inch Advantage Performance Upgrade Package to Reduce Fuel Burn and Operating Costs. 2010. Available from: www.pw.utc.com/Press/Story/20100205-0100/2010/Commercial%20Engines12.12.2017 [12 December 2017]

[72] World Bank. Annex D: Operating Cost. Available from: pubdocs.worldbank.org/en/777021436899472389/Air-Transport-Air-Cargo-AD.pdf17.11.2017 [17 November 2017]

[73] Goold I. Out with the old, in with the new. 2011. Available from: www.elal.com/en/ELALTech/Documents/Press_Releases/747.pdf15.11.2017 [15 November 2017]

[74] Strampfer P. First Austrian Airlines Embraer jet took off today for inaugural flight to Stuttgart. 2016. Available from: www.lufthansa.com/mediapool/pdf/84/media_82658284.pdf12.10.2017 [12 October 2017]

[75] KLM. KLM first European carrier to operate the EMBRAER 175 with enhancements. 2016. Available from: news.klm.com/klm-first-european-carrier-to-operate-the-embraer-175-with-enhancements/01.12.2018 [01 December 2018]

[76] Hemmerdinger J. Embraer details E-Jet efficiency improvements. 2013. Available from: www.flightglobal.com/news/articles/embraer-details-e-jet-efficiency-improvements-390504/10.12.2018. [10 December 2018]

[77] Sheppard I. LCC growth forcing European regionals to buy E-jets. 2011. Available from: www.ainonline.com/aviation-news/aviation-international-news/2011-12-01/lcc-growth-forcing-european-regionals-buy-e-jets01.12.2017 [01 December 2017]

[78] Embraer Commercial Aviation. Embraer: Leadership in regional aviation. 2013. Available from: doczz.net/doc/1157967/embraer--leadership-in-regional-aviation01.02.2018 [01 February 2018]

[79] Clark P. Air New Zealand adds Armonia interior to ATR 72-500s. 2013. Available from: www.flightglobal.com/news/articles/air-new-zealand-adds-armonia-interior-to-atr-72-500s-391920/03.11.2017 [03 November 2017]

[80] Thomson K, Schulze ET. Delivering fuel and emissions savings for the 777. 2009. Available from: www.boeing.com/commercial/aeromagazine/articles/qtr_03_09/pdfs/AERO_Q309_article02.pdf10/05/2016 [10 May 2016]

[81] Norris G. Mind the Gap: Engine and aero improvements are underway for Boeing's 777 in a run-up to 777X. 2015. Available from: iatdh.isti.ir/uploads/aviation_week.pdf?u=aviation+week+space+22.10.2018 [22 October 2018]

[82] Flightglobal. Engine evolution: HOW HAVE 737, A320, 777 AND A330 KEPT SELLING GRONGLY FOR SO MANY YEARS? 2015. Available from: www.flightglobal.com/sponsored/engine-alliance/13/10/2017 [13 October 2017]

[83] Aircraft Commerce. Replacement options for 80- to 150-seat jets. 2006. Available from: www.aircraft-commerce.com/__tools/_viewers/assetViewer.asp?id=26472532&wID=1645880429.01.2018 [29 January 2018]

[84] European Environment Agency. Emissions Inventory Guidebook. 2002. Available from: www.eea.europa.eu/publications/EMEPCORINAIR3/B851vs2.4.pdf/at_download/file19.01.2017 [19 January 2017]

[85] Flightglobal. ANALYSIS: After three years in service, how is 787 performing? 2014. Available from: www.flightglobal.com/news/articles/analysis-after-three-years-in-service-how-is-787-performing-405814/04.04.2017 [04 April 2017]

[86] Boeing Commercial Airplane Group. 787 Dreamliner By Design: The 787 Family: A Benchmark in Fuel Efficiency. 2017. Available from: www.boeing.com/commercial/787/by-design/04.04.2017 [04 April 2017]

[87] Derber A. The road to number TEN. 2017. Available from: www.mro-network.com/manufacturing-distribution/road-number-ten13.04.2018 [13 April 2018]

[88] Polek G. Boeing readies delivery of first 787-10 to Singapore Airlines. 2018. Available from: www.ainonline.com/aviation-news/air-transport/2018-02-01/boeing-readies-delivery-first-787-10-singapore-airlines13.04.2018 [13 April 2018]

[89] Euromoney Institutional Investor PLC. Air Investor Guide. 2007. Available from: airfinancejournal.com/docs/AFJ%20AirInvestor.pdf13/01/2017 [13 January 2017]

[90] GE Aviation. The GEnx Commercial Aircraft Engine: The GEnx engine delivers proven performance for the Boeing 787 Dreamliner and Boeing 747-8. 2017. Available from: www.geaviation.com/commercial/engines/genx-engine12.01.2018 [12 January 2018]

[91] Air Transportation Action Group. Aviation Climate Solutions. 2015

[92] Stonestreet K. Market & Commercial Overview. January 2015

[93] Norris G. Trent XWB-84EP upgrade leverages A350-1000 engine advances. 2016. Available from: aviationweek.com/singapore-airshow-2016/trent-xwb-84ep-upgrade-leverages-a350-1000-engine-advances25.10.2017 [25 October 2017]

[94] Fehrm B. Airbus increases A350-900 range to 8100nm. 2016. Available from: leehamnews.com/2016/03/30/airbus-increases-a350-900-range-8100nm/01.02.2018 [01 February 2018]

[95] Airbus. A320 Family Facts & Figures. 2016. Available from: www.airbus.com/presscentre/corporate-information/key-documents/?eID=maglisting_push&tx_maglisting_pi1%5BdocID%5D=10560110/01/2017 [10 January 2017]

[96] Polek G. Embraer reports E190-E2 spec improvements. 2018. Available from: www.ainonline.com/aviation-news/air-transport/2018-01-19/embraer-reports-e190-e2-spec-improvements16.02.2018 [16 February 2018]

[97] Baldwin M-A. The E2: Embraer's next generation of E-jets. 2013. Available from: issuu.com/afmmagazine/docs/afm-8501.01.2018 [01 January 2018]

[98] Salgado N. Current and future aircraft technologies. 2014. Available from: www.icao.int/Meetings/EnvironmentalWorkshops/Documents/2014-GreenTechnology/2_Salgado_Embraer.pdf12/04/2018 [12 April 2018]

Chapter 9

Climate Warming and Effects on Aviation

Diandong Ren and Lance M. Leslie

Abstract

The greatest concerns of the aviation industry under a warming climate possibly are the following two questions: first, what are the consequences for maximum payloads? and second, will changed air properties (density, temperature and viscosity) affect fuel efficiency? Here, the effects of climate warming on maximum payload and fuel efficiency are examined using atmospheric parameters from 27 climate models. Historical (20th century) climate simulations credibly reproduce the reanalysis period (1950–2015) of near-surface air density (NSAD). Lower NSAD is a first-order global signal continuing into the future. The NSAD reduction impact on MTOW could be ~1% over the busy North Atlantic Corridor (NAC), and also varies among aircraft. Furthermore, for the standard 7-stage flight profile, negative effects of warming on fuel efficiency affect civil aviation. The cruising stage consumes most aviation fuel, and as cruising altitude coincides with the tropopause, the tropopause structure in a warming climate supports the conclusions drawn here. Tropopause temperature changes cause only ~0.08% reduction in thermal efficiency. The net effect on total efficiency is smaller because of improved mechanical efficiency. Work required for a commercial aircraft increases in a warmer climate due to elevated tropopause altitude and increased air drag. The latter outweigh the former by almost an order of magnitude, for international flights.

Keywords: tropopause, civil aviation industry, climate warming, aircraft payload, fuel efficiency, GCMs, climate change adaptation

1. Introduction

Aviation is probably the most reliable means of disaster response and relief for most large-scale natural disasters. For example, it is unrealistic to maintain or repair road or rail connections across areas affected by earthquakes, floods, landslides, storms, or wildfires, to transport relief and aid to those affected. Thus, adaptation and risk management must pay particular attention to the strengthening of aviation infrastructure to guarantee robust and sustainable relief. By providing a perspective on the impacts on aviation of anticipated changed atmospheric conditions over the near future, this research addresses the adaptation of aviation transport to climate change. The greatest concerns of the aviation industry under a warming climate possibly are the following two questions: how will the maximum payload be affected by the warmer and lighter lower layer atmosphere? and, during the journey, will the changed ambient air properties (density, temperature and viscosity) affect the engine performance? Anyway, all current aviation engines are breathing thermal engines. The first part of this chapter focuses on the maximum payload, whereas the second part concentrates on the effects on the efficiency and fuel consumption of the thermal engines. Commercial airliners provide an environment-friendly express means of cargo transport and personnel travel (Section 7.4.1.2 of IPCC AR5 [1]). Possible effects of aviation on atmospheric components and climate already have been studied in detail [2–7]. Conversely, the the effects of climate warming on aviation have not yet been extensively studied. In Section 1 of this chapter, climate warming effects on aviation payload are investigated, based on the fact that air density is proportional to the maximum take-off weight (MTOW) for an aircraft, irrespective of the design (fixed wing or helicopters; jets or propellers). Aircraft are air-lifted and the MTOW can be expressed in a generic form as

$$MTOW = \mathrm{M}\rho_a \left|\vec{V}_0\right|^2, \tag{1}$$

where M is airplane mechanical properties such as wing span area, attack angle, and fuselage-wing interaction, ρ_a is air density, and $\vec{V}_0$ is aircraft taking-off speed (for helicopters and other rotorcrafts, the angular speed of the rotor blades). If taking-off speed also is deemed as aircraft property, then air density is the sole environmental factor that is directly proportional to $MTOW$. In this study, near-surface (airport elevation) air density variations are examined over the twentieth and twenty-first centuries. For period with reanalysis data (1950–present), the loyalty of air density simulated by 27 climate models (**Table 1**) to reality also is examined. Based on the analyses, the decrease of maximum payload is examined, and the inter-model spread of the uncertainty assessed up to 2100. The largest uncertainty in the degree of warming resides with the industrial emission of GHGs and other pollutants in the atmosphere to which climate is sensitive to the extra radiative forcing, because modern climate has a clear footprint of human activity [2, 8]. The future state of climate would depend crucially on what emission controls nations chose to impose. Emission scenarios (ESs) describe future release into the atmosphere of GHGs, aerosols, and other pollutants. ESs and other boundary conditions are inputs to climate models. In the most recent Intergovernmental Panel on Climate Change (IPCC) assessment report (AR5), the driving scenario is in the form of representative concentration pathways (RCPs). In this study, the climate model outputs under high scenario RCP 8.5 (meaning that rising radiative forcing pathway leads to 8.5 W/m^2 heating effects in 2100) are used.

The second theme of this chapter is on aviation fuel efficiency. According to FAA regulations, the flight profile consists the seven stages (A-G from taxi out till taxi in). To estimate the extra work that needs to be performed, along flight route integration is the exact approach. Because the commercial data on flight logs are not available for us, we have to make some assumptions according to the carrier aircraft and the routes, which are readily available online (e.g., from those websites selling air tickets). Unlike the issue with maximum payload, where only the airport level air density plays the decisive role, temperature, air density, and winds all matter in the fuel efficiency issue. There sure exist apparent canceling effects among them as well. In addition, the tropopause's elevation will fluctuate as climate warms; this involves extra potential energy cost in case the aircraft still cruise in the coldest (hence the most favorable for the thermal engine) and most clear level of the Earth's atmosphere. This likely is the case since the cruise stage is the most fuel-consuming stage of the flight (although the rate of fuel burning is only a half of

Model Name	Institution	Horizontal resolution
ACCESS1-0	Commonwealth Scientific and Industrial Research Organization (CSIRO)/Bureau of Meteorology (BoM)	192×145
ACCESS1-3	CSIRO/BoM	192×145
CCSM4	National Center for Atmospheric Research	288×192
CanESM2	Canadian Center for Climate Modelling and Analysis	128×64
CNRM-CM5	Centre National de Recherches Meteorologiques	256×128
CESM1-BGC	National Center for Atmospheric Research	288×192
CESM1-CAM5	National Center for Atmospheric Research	288×192
FGOALS-g2	Institute of Atmospheric Physics, Chinese Academy of Sciences	128×60
GFDL-CM3	Geophysical Fluid Dynamics Laboratory	144×90
GFDL-ESM-2G	Geophysical Fluid Dynamics Laboratory	144×90
GFDL-ESM-2M	Geophysical Fluid Dynamics Laboratory	144×90
GISS-E2-H	NASA/Goddard Institute for Space Studies (GISS)	144×90
GISS-E2-R	NASA/GISS	144×90
HadGEM2_AO	National Institute of Meteorological Research, Korea Meteorological Administration (S. Korea)	192×145
HadGEM2_CC	Met Office Hadley Centre	192×145
HadGEM2_ES	Met Office Hadley Centre	192×145
INM	Institute of Numerical mathematics, Russian Academy of Sciences	180×120
IPSL-CM5A-LR	Institut Pierre Simon Laplace (France)	96×96
IPSL-CM5A-MR	Institut Pierre Simon Laplace (France)	144×143
IPSL-CM5B-LR	Institut Pierre Simon Laplace (France)	96×96
MIROC5	Atmosphere and Ocean Research Institute, The Tokyo University	256×128
MIROC-ESM	Japan Agency for Marine-Earth Science and Technology	128×64
MRI-CGCM3	Meteorological Research Institute	320×160
MPI-ESM-LR	Max Plank Institute (MPI) for Meteorology	128×64
MPI-ESM-MR	Max Plank Institute (MPI) for Meteorology	192×96
NorESM1-M	Bjerknes Center for Climate research, Norwegian Meteorological Institute	144×96
NorESM1-ME		144×96 global grids

Table 1.
Twenty-seven GCM models used in this study.

the ascending stage). So, a complete consideration of the issue involves the along flight route integration plus the potential energy changes of the entire aircraft. Discussion bifurcates in the following sections, focusing, respectively, on the effects on maximum payload (Section 2) and fuel efficiency (Section 3). This study should inspire further investigation into how climate and environmental changes influence the civil aviation sector of industry.

2. Adverse effects on maximum payload from a warmer climate

2.1 Methods and data

Air density is derivable from air pressure, temperature, and humidity [9, 10]:

$$\rho_a = P/[R_d T(1 + 0.608q)] \tag{2}$$

where P is air pressure (Pa), R_d is dry air gas constant (~287 J/kg/K), T is absolute air temperature (K), and q is specific humidity (g/g). To apply Eq. (2) to near-surface level, atmospheric fields of pressure (Ps), temperature (Ts), and specific humidity qs at ground level (subscript "s" means surface) are required. These parameters fortunately are primary outputs from the coupled model intercomparison project (CMIP, e.g., https://cmip.ucar.edu/; Ref. [11]). The monthly climate model outputs are obtained from the IPCC Deutsches

Klimarechenzentrum (DKRZ) Data Distribution Centre (http://www.ipcc-data.org/sim/gcm_monthly/AR5/Reference-Archive.html). For models providing multiple perturbation runs, only *r1i1p1* runs are used. To examine whether the historical runs from the climate models are close to reality in their simulated air density, NCEP/NCAR reanalyses are used as observations. The monthly NCEP/NCAR reanalysis [12] data are obtained from the Earth System Research Laboratory website: http://www.esrl.noaa.gov/psd/data/gridded/data.ncep.reanalysis.pressure.html. Specific humidity provided by reanalyses can be converted into specific humidity before applying Eq. (2).

2.2 Results and discussion

From Eq. (2), two factors cause air density fluctuations: temperature changes and mixing in tracers of different molecular weight to the average molecular weight of the air. Earth atmosphere is dominated by the "dry" inactive components (N_2, O_2, CO_2, etc.). With heat intake, the primary response is expanding in volume and, subsequently, an increased mass center. During the past half century, on average, there is a 30 m lift of the mass center, indicating that the mass is now distributed in a thicker (larger depth) layer (thus reduced density at lower levels). Heating caused expansion is just one effective means that decreases air density throughout the entire tropospheric atmosphere. In-taking of lighter molecules

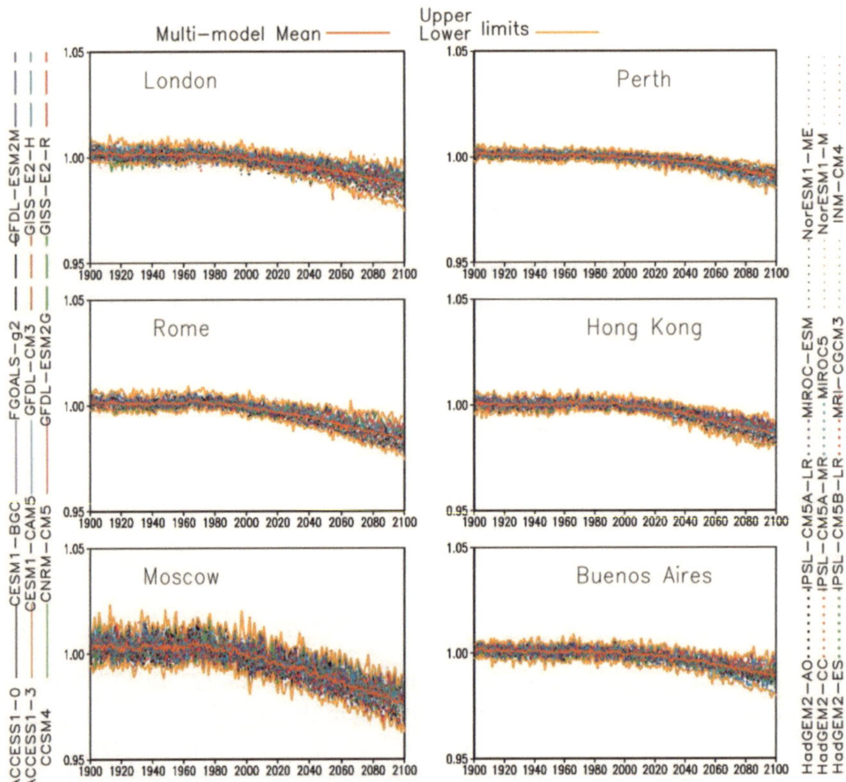

Figure 1.
Changes in near-surface air density for the period 1900–2100 as simulated by 24 climate models over six selected world airports. Near-surface air density estimated from NCEP/NCAR reanalyses is shown as red bold lines. The upper and lower bounds among the 24 climate models are shown as brown lines. The analyses were performed on monthly data and averaged to annual values (actual plotted). As with all land-ocean-ice sheet fully coupled climate model outputs, the exact timing is hard to pinpoint. Only the statistical properties and long-term averages would resemble reality.

(H_2O has smaller molar mass than N_2 and O_2, the dominant constituents of dry air) is another effective way of reducing air density. Although from Clausius-Clapeyron equation [13] warm air has more capacity of holding moisture, it still is debatable whether earth atmosphere actually gains mass, because the hydrological cycle also tends to intensify [14–16], through facilitating interhemispherical moisture exchange [17] and destabilizing local stratification profile [15, 16]. If precipitation increases more than evaporation, there still is a net mass loss for atmosphere. Interestingly, all existing reanalysis datasets show no statistically significant changes in global total air mass during their respective reanalyses period. This implies that the net water vapor input into atmosphere is globally delicately balanced between geographic regions.

Applying the formula as in Eq. (2) to climate model-simulated (under RCP 8.5, a strong emission scenario) near-surface pressure, temperature, and humidity, near-surface air density is estimated over the globe. The same formula also is used on the NCEP/NCAR reanalysis data. Density variations over 1900–2100 for six global airports are shown in **Figure 1**, as representatives. All 27 climate models unanimously indicate that all six locations experienced salient density decreases. Significant inter-model spread exists but started well before the year 1900 and should be ascribed to model systematic biases/drifts. For each climate model, the amount of density decrease easily exceeds the natural interannual variability magnitude. Geographically, high-latitude regions (e.g., Moscow) have larger

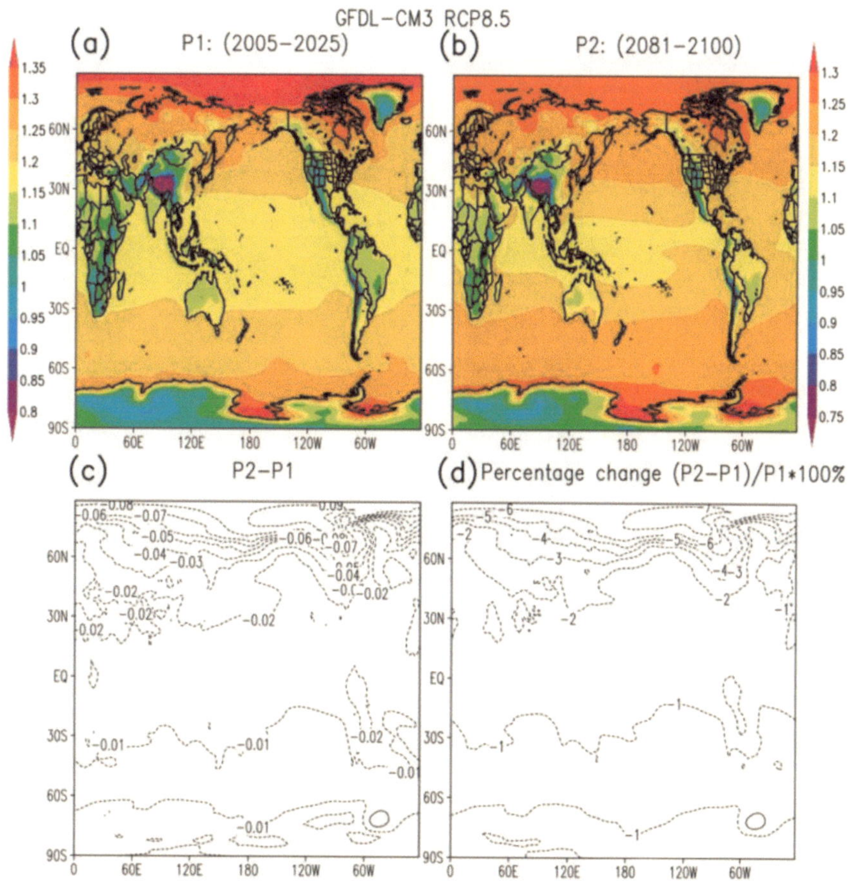

Figure 2.
GFDL-CM3-simulated near-surface air density (kg/m³) averaged over two periods: (2005–2025) (a) and (2081–2100) (b), under RCP 8.5 scenario assumption. The density differences between these two periods are shown in (c), with corresponding percentage changes shown in (d).

interannual density variability but also generally experiences larger density decreases over the simulated 200 years. The linear trends of decrease estimated based on the reanalyses are close to model simulation. All 27 climate models show high degree of consensus in the simulated air density changes (e.g., **Figure 2** for GFDL-CM3).

Air density changes are a gradual process over the years. To quantitatively examine if the density changes were statistically significant, a *t*-test was performed over the two 20-year annual density time series (2005–2025 and 2081–2100). In **Figure 3**, the *t*-value (right panels) and corresponding *P*-values (left panels, for a DoF of 38) are presented (six climate models are shown to demonstrate this). Under the RCP 8.5 scenario, most global regions pass the 95% confidence interval. The signal-to-noise ratio is low only for very limited oceanic regions off the southern tip of Greenland and some sectors of the Southern Ocean. The oceanic region off southern Greenland collocates with the deep-water formation region of the North Atlantic meridional overturning circulation (MOC). Further investigation indicates that the lack of a decrease in density is due to a weakened MOC that places less moisture into the atmosphere. Regionally, the air becomes drier and heavier. On the other hand, warming reduces the air density. These two canceling factors make the net reduction in density statistically insignificant or, when the moistening effect wins out, may even increase the air density. For this study, the oceanic regions that

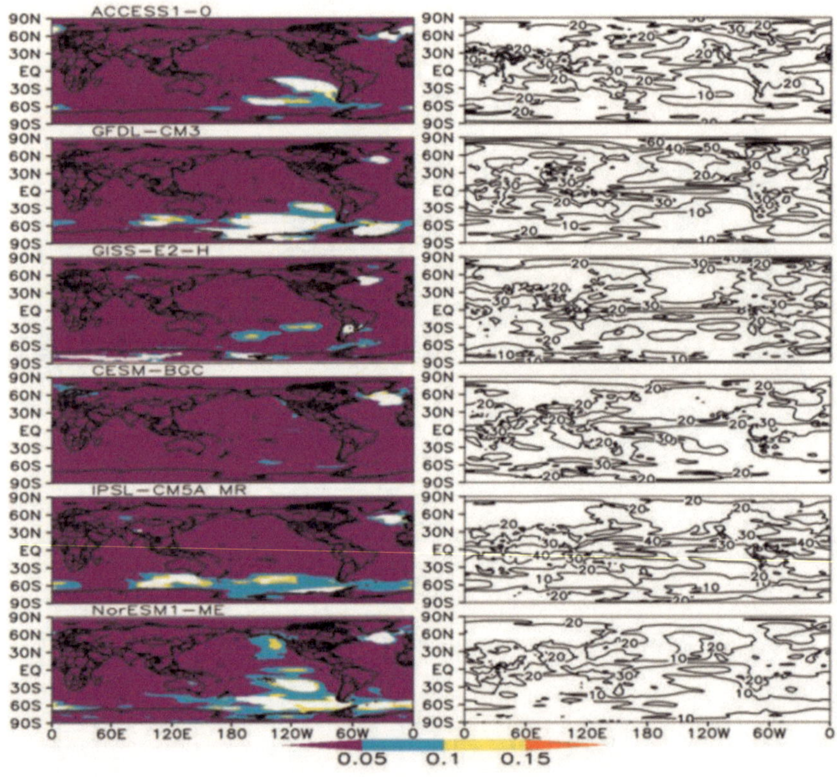

Figure 3.
A t-test was performed for near-surface air density (simulated under RCP 8.5 scenario) annual time series over two periods (2005–2025) and (2080–2100). Sophisticated modern climate models show consensus on the geographical patterns of the P-value (left panels) and t-values (right panels). Except for some oceanic regions off the southern tip of Greenland and some portions of the Southern Oceans, most global regions passed the 95% confidence interval, for a DoF of 38. Tropical regions, especially over the intertropical convergence zone (ITCZ), experienced the most significant density decreases. The oceanic region off southern Greenland collocates with the region of deep-water formation of the North Atlantic meridional overturning circulation (MOC).

did not pass the *t*-test can be disregarded, because a very limited number of airports are marine-based. Unlike the percentage changes, the tropical region, especially the intertropical convergence zone (ITCZ, an area of low pressure and convergence of trade winds), had the lowest *P*-value, meaning that the changes over the region are most likely to be statistically significant. From Eq. (2), air density is co-controlled by temperature change and vapor content change. The temperature changes over tropical regions are smaller than over the high-latitude regions. What make the changes over tropical regions statistically more significant are the relatively small changes in air density, for all time scales.

Although air density values simulated by the 27 climate models, when compared with those derived from the NCEP/NCAR reanalysis data, have systematic biases, the linear trends derived from models agree very well with the reanalyses. This indicates that for estimating payload decreases as the climate warms, the density time series can be normalized by their average value over a control period, say 1900–1920. For a specific model, differences in the values of the normalized density time series from unity are the percentage reductions of NSAD and MTOW. If one further assumes an invariant unavoidable load (weight of an empty airplane), the decrease of MTOW also is the decrease of maximum payload. Air density changes estimated from all climate models were interpolated to the same spatial resolution as MRI-CGCM3. **Figure 4** shows the MTOW changes between the two 20-year periods (2005–2025 and 2080–2100). Globally, the changes can reach 5% reduction for some high-latitude and high-elevation airports. For the busy North Atlantic Corridor (NAC), the reduction generally is greater than 1%. This has important economic significance. For the Boeing 747-400, this means a net load reduction of about 3969 kg (**Table 2**), approximately the passenger and luggage weight of ~25 passengers, or a ~6% reduction in its full passenger carrying capacity. Actual payload equivalence of a 1% reduction in MTOW for other types of aircraft is listed in **Table 2**. Because the ratio of unavoidable load to its maximum effective payload varies for different aircraft, the general reduction in net payload over NAC varies from 5 to 8.3% for all the aircraft types considered. Some Northern Hemisphere

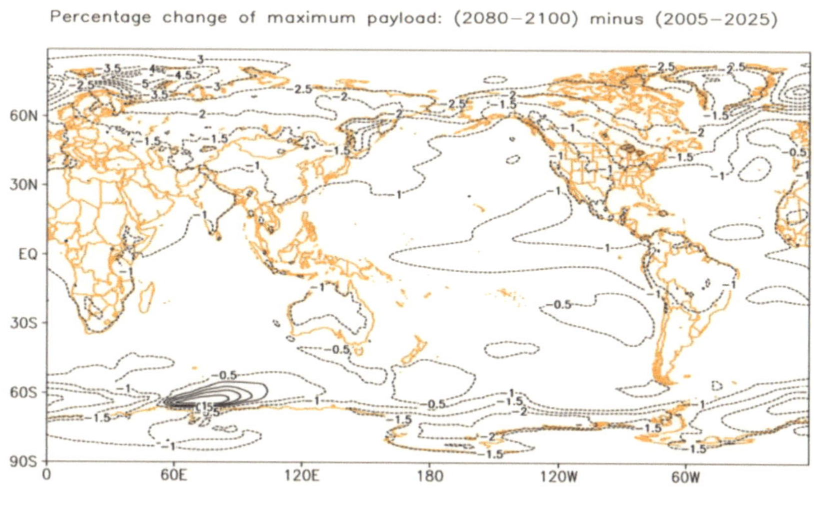

Figure 4.
The estimated ensemble mean decrease in aircraft maximum takeoff total weight (MTOW), as a percentage, based on air density changes simulated by 27 climate models under the RCP 8.5 scenario. The near-surface density from each climate model was normalized by its mean value over 2005–2025 (the control period). Then a bilinear interpolation scheme was used to interpolate on to MRI-CGCM3's horizontal resolution. An ensemble average was taken over the climate models. The reduction can reach 5% over Northern Europe. For the North Atlantic Corridor (NAC), a ~1% reduction in MTOW was reached during the 75-year span.

Aircraft Type	1% MTOW [kg]	Aircraft Type	1% MTOW [kg]	Aircraft Type	1% MTOW [kg]	Aircraft Type
Bombardier (Bom) CS300	653.2	Antonov An-124-100M	4050.6(1.7%)	Douglas DC-8-61	1474.2	McDonnell Douglas MD-9(30
Vickers Viscount 800	303.9(3%)	Lockheed L-1011-200	2113.7	Handley Page Hermes	390.1	de Havilland (dH Comet 1
Airbus A330-200	2300(1.97%)	B767-400ER	2041.2	B727-100	771.1	Bom CS100
Tupolev Tu-154M	1043.3	Bom CRJ700	329.8	MD-83 (2.2%)	725.7	Ilyushin IL-96-30
A380-800	5443.1(1.9%)	Boeing 787-8	2279.3	B707-120B	1165.7	A318
Boeing B747-8	4477.0(2%)	A350-1000	3079.9	B707-320B	1513.2	B737-600
B747-400ER	4127.7	IL-86	2079.7	B720B	1061.4	B737-400
B747-400	3968.9	B767-300ER	1868.8	B727-200	838.2	B737-300
B747-200	3778.4	A300-600R	1719.1	HS Trident 2E	646.4	B717-200HGW
B747-300	3778.4	A300-600[7]	1632.9	Caravelle III	459.9	B737-500
A340-200	2535.1	B737-900ER	851.4	DC-6B	485.3	Embraer ERJ 145
A340-500	3719.5	A310-300[7]	1569.9	dH Comet 2	544.3	Embraer 190
B777-200LR	3474.5	A310-200	1416.6	Convair 880	875.4	Embraer 175
B747-100	3333.9	A400M	1409.8	ATR 42-500	186.0	Embraer 170
B777-300	2993.7	B757-300	1236.0	ATR 72-600	227.7	Bom CRJ900
MD-11	2732.9	B757-200	1156.7	DC-4	331.1	Bom Q400
Il-96M	2698.9	Tu-204SM	1043.3	DC-6	440.9	Saab 2000
A350-900	2680.7	A321-100	829.6	DC-6A	486.3	Bom CRJ200
A340-600	3674.1	the Concorde	1850.7	dH Comet 3	680.4	B717-200BGW
B777-300ER	3515.3	B767-300	1587.6	dH Comet 4	707.6	Fokker 100
B777F	3478.1	Vickers VC10	1518.6	dH Hercules	70.8	Avro RJ-85
B777-200	2472.1	B737-900	851.4	DC-7	553.4	A330-300
A350-800	2590.0	B737-800	790.2	DC-8-32	1406.1	B737-700
A340-300	2766.9	A319	640.0	DC-8-51	1251.9	An-225
B787-9	2508.4	A320-100	679.9	A380-800F	5896.7	B377

Table 2.
Actual payload equivalence of a 1% reduction in near-surface air density (NSAD). Percentages of maximum payload equivalence are shown in parenthesis.

high latitudes have a ~5% decrease in NSAD or MTOW. Considering that a 1% reduction in MTOW corresponds to a ~2% (for larger aircraft such as a Boeing 747-800 or an An-24) to ~3.6% (for small aircrafts such as an Embraer ERJ-145) reduction in effective payload, the ~5% reduction in MTOW means a ~8.5–19% reduction in effective payload year-round. As we stated earlier, at the costs of extra maintenance, aircraft still can operate with the manufacturer-labeled MTOW, which is lower than MTOW, under the unfavorable condition of warming. There may be no apparent passenger or cargo reduction. However, there will be hidden extra costs from a warming atmosphere.

2.3 Discussion

Based on the diagnosis of stresses (and forces) exerted on aircraft, a suitable invariant entity was identified for investigating climate change effects on aviation payload. Assuming no changes in technical aspects of aircraft and no changes to FAA regulations on takeoff performance, near-surface air density is the single most significant atmospheric parameter. Reanalyses data indicated clearly that the Earth's atmosphere had expanded in volume in the past half century.

Consequently, the near-surface air density experienced significant decreases globally. The 27 climate models showed a high level of consensus in simulated near-surface air density variations. The ensemble mean of their twenty-first century simulations in NSAD trends was used to examine future reduction to effective payload. In line with Ref. [18], our study aimed to illustrate the potential for rising temperatures to influence weight restriction at takeoff stage. All technical aspects as commented on by Ref. [19] were assumed to be invariants during the analyses period. The simple fact that during extreme hot weather in summertime cargo airplanes have to reduce the effective payload indicates the validity of such analyses. The difference found with seasonal cycle is that these superimposed effects work persistently year-round and there is no easy way to circumvent or ameliorate them. We, however, agree with Ref. [19] that aviation industry still has technical room to cope with the detrimental effects from climate warming, perhaps at the extra costs of maintenance, passenger comfort, and may even require relaxation of aviation code.

3. Adverse effects on civil fuel efficiency from a warmer climate

Aviation fuel efficiency is underpinning recent contest between aviation engine makers—using higher bypass engines and improving higher fuel-burning temperature. Aside from technical challenges, further improvements in fuel-burning efficiency may also have safety consequences. In the following discussion, a normal seven-stage flight profile (these are A, start and taxi to runway; B, takeoff and initial climbing; C, climbing to cruising altitude; D, en route cruising; E, descent; F, approach (includes ~8-minute holding at ~1500 ft. approach and landing); and G, taxi to docking) is considered (**Figure 5**). **Figure 5** also shows the typical fuel-burning rates at different stages of a commercial airplane engine.

In this subsection, we start from the theoretical expression of the total work an aircraft needs to perform from the origin airport to the destination airport. Another aspect of the fuel efficiency issue is related to the second law of thermodynamics. All airplane engines are thermal engine. Increased environmental temperature always is detrimental for thermal efficiency. This is directly related to the fuel costs of civil aviation. With changed atmospheric thermal structure, the aircraft's mechanical efficiency may also vary; suppose the same FAA regulation is in position. All components that are sensitive to climate change are investigated and quantitatively from climate model simulations under the RCP 4.5 emission scenario—a more likely scenario.

3.1 Methods and data

During different stages of flying, the force balance situation on an aircraft is different. At the takeoff and climbing stages, there are vertical and forward accelerations. The vertical component of thrust aids the lift in overwhelming gravity. Similarly, the horizontal component of thrust also overwhelms drag. At cruising

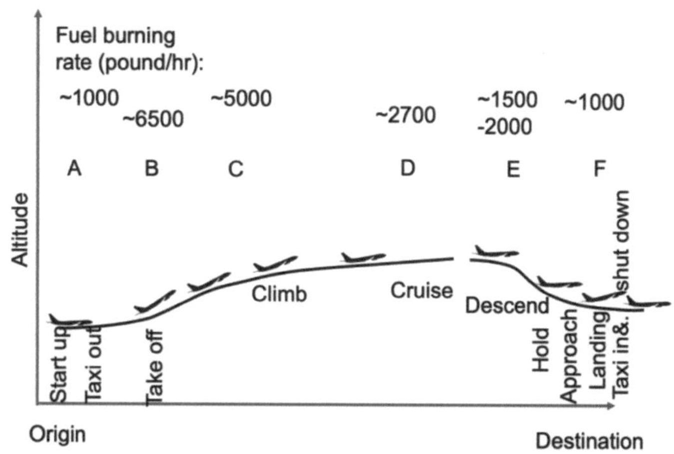

Figure 5.
Typical flight profile of an aircraft and the fuel-burning rate in each stage. Except the cruising stage, other six stages last from 10 to 40 minutes only. In all, cruising stage is the most fuel-consuming stage.

stage, thrust is reduced mainly to counteract drag, and the weight is balanced primarily by lift. Inevitable work done to the aircraft involves lifting it to the cruise elevation. The potential energy cannot be reclaimed at descending stage, unlike electronics cars. This portion of energy only is sensitive to warming when tropopause height changes as climate warms. The far larger term in energy cost would be that used to overcome drag. While it is apparent that drag is proportional to fuel cost, the picture for how total drag is affected by climate change is more sophisticated, because of the multiple sources involved. Classifying the many drag terms into pressure drag (e.g., induced drag, wave drag, and form drag) and skin friction drag (second term on right-hand side of Eq. (3)) is convenient because the pressure drags tend to be proportionally affected by air temperature and density changes. For example, for a specific design, the effects from environmental air on induced drag and net lift are usually proportional. Thus, the changes in skin friction are decisive for the sign of extra drag on top of total drag stress. To separate out climate change effects on aviation, it is assumed that there is no technological advance in design of subsonic aircrafts used for commercial airliners during the timespan of consideration:

$$F_d = \iint_S (p - p^*)\hat{n}\hat{i}dA + \iint_S \tau\hat{t}\hat{i}dA \tag{3}$$

where F_d is the total drag, $\hat{n}$ and $\hat{t}$ are, respectively, unit vectors in the direction perpendicular and parallel to the local surface element (dA), p is pressure, and $\hat{i}$ is the flow (drag) direction (align with the aircraft trajectory in **Figure 5**). For a specific type of airplane, the second term on the right-hand side of Eq. (3) can be parameterized as $p^*Sc_1R_e^{c_2}$, where S is wing area and R_e is Reynolds' number. Coefficients c_1 ($\sim$0.074 for well-painted un-dented surfaces) and c_2 (approximately$-$0.2) are aircraft dependent. The drag coefficient is inversely related to the Reynolds number. Increased flow speed tends to increase R_e, while increased temperature, with consequently increased dynamic viscosity, tends to reduce R_e. Without resorting to strict model calculation, it is difficult to estimate accurately the net effect from climate warming to total drag.

To have an estimate of the effects by the end of the twenty-first century, we followed a line-by-line analysis of available commercial airliners. For this purpose, online commercial ticketing databases are browsed for available flights among global airports. Non-direct flights are decomposed to several "direct flights" in a row. An annual, global, direct flight database is thus archived for this research. To estimate the total annual fuel consumption, we follow a line-by-line adding method that considers all available (in operation as of 2010) commercial airliners and their scheduled flights. The integration is along the flying trajectory. There are all sorts of alliances and partnerships between the commercial airliners. A trip involving multiple stops is likely carried out by different airliners in collaboration. For example, between Beijing and Singapore, there are 14 companies having such a transportation service at sub-weekly frequency. Asiana and Air China, for example, have a service to take passengers to Seoul first before heading to Singapore. Cathay Pacific and Thai Airlines stop, respectively, in Hong Kong and Bangkok. Xiamen Airlines even make two stops in between (Beijing → Zhoushan → Xiamen → Singapore). To eliminate possible recounting of the flying legs, only direct flights (each involves one takeoff, cruise, and one landing) between airports are analyzed. In the above case between Beijing and Singapore, there are only five such daily flights, from Air China (A975 and A976, Airbus 330 s) and Singapore Airlines (SA801, 805, Boeing 777 s as carrier, and SA 807, an Airbus 380-800). Connecting flights from the same airline or from several partner airlines are considered to be several connected direct flights, with distinct flight profile legs and usually carried out using different types of aircraft. For example, the Xiamen Airline schedule from Beijing to Singapore is looked upon as a direct flight from Beijing to Zhoushan, followed by a direct flight from Zhoushan to Xiamen, and another direct flight from Xiamen to Singapore. In this specific case, the same types of aircrafts are used. However, for intercontinental flights, usually different types of aircrafts are involved and intercontinental legs cruise at a higher elevation than the domestic legs of flights.

In the estimation of fuel efficiency change by the end of this century, atmospheric parameters (i.e., air temperature and humidity) from multiple climate models (all under RCP 8.5 scenario) are used to drive expressions (Eqs. (3)–(5)), weighted by airplane-specific aerodynamic parameters. Ensemble averages are taken after the along trajectory integrations driven, respectively, by all climate models (**Table 1**). The climate model outputs are obtained from the IPCC Deutsches Klimarechenzentrum (DKRZ) Data Distribution Centre (http://www.ipcc-data.org/sim/gcm_monthly/AR5/Reference-Archive.html). For models providing multiple perturbation runs, only $r1i1p1$ runs are used. There still is quite a large uncertainty with emission scenario. The results presented here thus should not be taken too literally. Instead, its values primarily are quantitatively accurate.

3.2 Results

From the discussion in Section 3.1, we see that the total energy an aircraft needs to perform is the one overcoming the drag force (Eq. (3)) and the one overcoming gravity to the cruising altitude. The drag forces do work all stages taking off and before landing (all the suspension stages), whereas the potential energy increases only during the taking off and climbing to the cruising altitude (usually tropopause elevation for best visibility and thermal efficiency—to be discussed soon). Because the cruising stage is of very different lengths, in the following discussion, we estimate the percentage change in energy (fuel) costs relative to each stage in the A-G profile against their respective values (e.g., changes in fuel cost in each flight stage, rather than vaguely relative to the total seven stages).

3.2.1 Fluctuation of the tropopause height

From **Figure 6**, tropopause has apparent latitudinal distribution: reaching lower pressure (higher altitudes) at the tropical region and drops to higher pressure levels at the polar regions. As climate warms, tropopause was lifted to higher elevations (**Figures 6b** and **7a** and **b**), except very localized regions around the South Pole. This is in agreement with Refs. [20, 21]. Different emission scenarios differ primarily in magnitudes, with decreasing regions totally disappeared for the strong emission scenario RCP 8.5. For the bustling North Atlantic Corridor (NAC, 305–350E; 30–60N), not only the trend but even the differences between the two scenarios (**Figure 7c**) pass a t-test with 95% confidence interval. The tropopause altitude increase rate reaches 6 m/year for a weak emission scenario (RCP 4.5). This is about 4% increase in the fuel cost at the ascending stage for normal commercial flights.

In the ascending stage, (in the vertical direction) the aircraft not only overcomes gravity, but it also experiences drag (both terms in Eq. (3)). As a result, the ascending at the lower altitudes is more fuel-consuming, because of the higher air density. As a result of this fact, the elevated tropopause elevation is only an increase of less than 0.2% in fuel costs for long-range international flights. Except for very short-range flights, the cruising stage is the most fuel-consuming stage. Factors

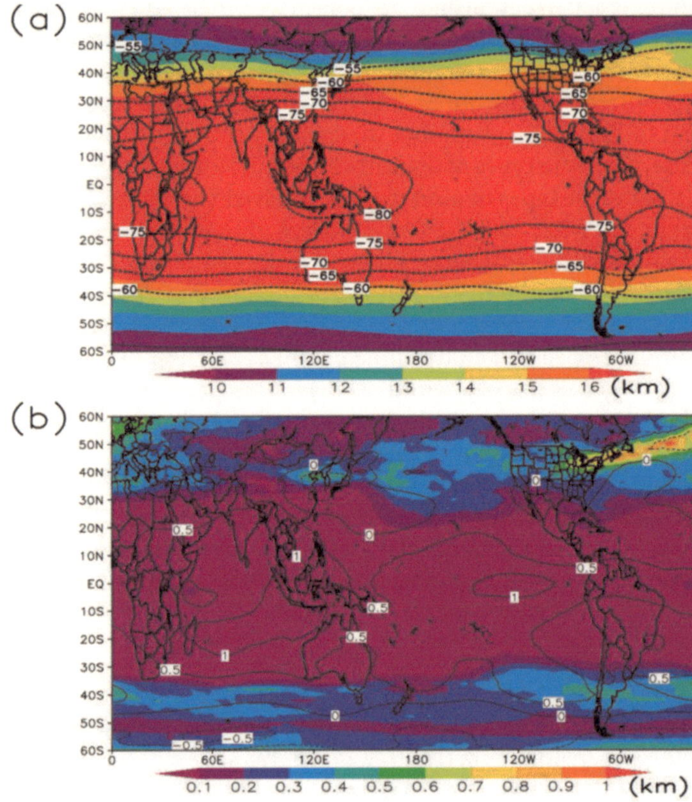

Figure 6.
GFDL2.1 simulated tropopause height (shades in (a), in km) and tropopause temperature (contour lines in (a)) during a control period (1980–2000). The projected differences between (2080–2100) and the control period, under the RCP 8.5 emission scenario, are shown in (b). The increases in tropopause height are a global phenomenon. For most areas, tropopause temperatures also increase.

affecting the cruising stage needed to be examined to have an estimate of the fuel efficiency issue of climate warming.

3.2.2 Thermal, mechanical, and total efficiency

Aircraft engines are breathing thermal engines. That is, they use oxygen in the environmental air fanned into the burning chamber, rather than carrying the oxidizers (as rocket engines do) for burning the fuel. The working fluid is the high-temperature and thus high-pressure exhausts (gases resulting from burning of fuel plus other components in the inhaled air). As fuel and inhaled air are "locked" in the burning chamber moving with the aircraft, the overall efficiency (in providing thrust) is a multiplication of thermal efficiency and mechanical efficiency. Applying Newton's third law of motion (or momentum theory) $F\Delta t = m\Delta V$, it is straightforward to ascertain that the overall efficiency, η, is the multiplication of mechanical efficiency (η_M) and thermal efficiency (η_T), or $\eta = \eta_M \times \eta_T$ (e.g., Ref. [22]).

As aircraft engines are thermal engines, their thermal efficiency is adversely affected by environmental temperature rise. The second law of thermodynamics puts a fundamental limit on thermal efficiency (η_T):

$$\eta_T = \varepsilon(1 - T_C/T_H) \tag{4}$$

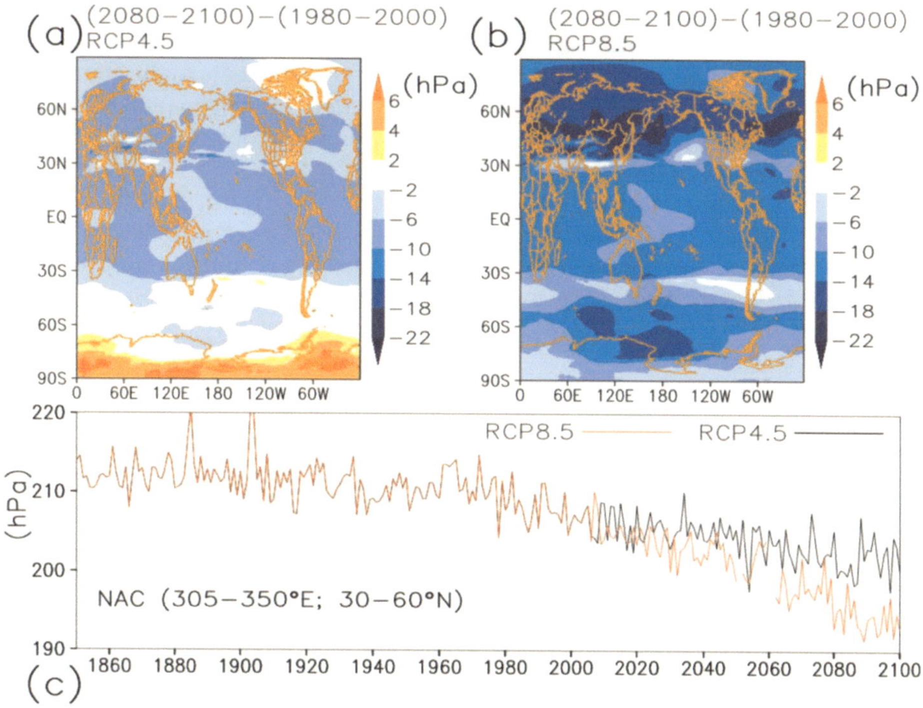

Figure 7.
GFDL2.1 simulated tropopause height changes (shades in (a) and (b), in hPa) between periods 1980–2000 and 2080–2100, represented in pressure levels. The area averaged time series over the NAC (305–350E; 30–60N) is shown in (c), for two emission scenarios (RCP 4.5 and RCP 8.5). The two differs only quantitatively.

where ε is a technical limiting factor ($\sim$0.57) indicating actual engines' closeness to ideal engine [9], T_H is the absolute temperature at which the heat enters the engine cycle (also called turbine entry temperatures, TETs), and T_c is the absolute temperature of the exhaust gases. T_c closely follows the environmental air temperature, T_a, with only a cooling technology-dependent constant difference. The efficiency of thermal engines increases with higher operating temperature and lower environmental temperature. There have been active efforts in improving TETs during the past 50 years (Section 7.4.1.2 of IPCC AR5: aviation and the global atmosphere). In this study, for simplicity, we assume that both T_H and ε are not going to improve in the projection period (between 2010 and 2100). T_c is the only variable being considered varying along the cruising route. For all commercial brands in operation, the TETs published during 2010–2014 are used. Except for several well-known engine types, most engine companies are very protective of actual engine data and operating conditions, although there is much discussion in the literature and also clues such as in EASA (and FAA) type of certificates (certificates for all of the engine types are publicly available), which have detailed listings of actual engine values in regions where engine measurements are made. The TETs can be deduced from the emission data. According to Eq. (4), a decrease in thermal efficiency for common commercial engines, GE90, RB211, LAEV2500, and Lyulka, in response to a 1 K increase in environmental temperature (T_c), are respectively, 5.62×10^{-4}, 5.88×10^{-4}, 5.07×10^{-4}, and 4.94×10^{-4}. The effects on engine thermal efficiency from climate change are transferred to the air temperature variations along the flying routes (legs). As the oxygen is inhaled from environmental air (common to breathing engines), and the density decrease due to climate warming does not result in incrementing oxygen concentration in the environmental air (as a matter of fact, if vapor content is considered, a decrease in oxygen concentration is expected), a natural consequence is that this may result in incomplete fuel oxidization, without technological improvements to the combustion system to increase the volumetric air inhaling rate. This detrimental effect on thrust production is not considered here but is apparently proportional to air density decrease.

As the airplane moves forward by ejecting exhausts backward, the way in which the kinetic energy (extracted from the fuel-burning chemical energy) is partitioned between aircraft and exhaust jet (i.e., used for pushing aircraft forward versus removed by the exhaust) is measured by the mechanical (propulsive) efficiency (η_M):

$$\eta_M = \frac{2}{1 + V_e/V_a} \tag{5}$$

where V_e is effective exhaust speed (jet speed relative to airplane) and the airplane speed, V_a, is relative to the ground. η_M reaches maximum when the jet exhaust is stationary relative to the ground (all extracted energy from fuel burning is used as thrust to push forward the aircraft). Here exhaust speed is retrieved from TETs and engine pressure ratios [23]. Equation (5) is derived in the inertial frame coordinates based on energy and momentum conservation. The commercial passenger aircrafts generally have effective jet speeds, V_e, within the range of 600–850 m/s. The lower the effective jet speed (i.e., close to the cruising speed), the more sensitive the mechanical efficiency is to airplane cruising speeds. Overall efficiency η is the multiplication of mechanical efficiency and thermal efficiency ($\eta = \eta_M \times \eta_T$). Factors lowering (enhancing) overall efficiency result in more (less) fuel cost.

In the global belt between 65°S and 70°N, which contains most of the trajectories of commercial flights, the tropopause temperature increased ~0.8–1.2°C over a 100-year period (the difference between (2080–2100) and (1980–2000)). For most commercial engines, thermal efficiency reduces only 0.06% during the cruising stage of the flight profile. Due to air density decrease, the mechanical efficiency is affected by warming the opposite way. As a result, the total efficiency was affected only by ~0.03%. The most significant effect from a warming flying environment is in fact from the increased air stickiness—the body drag acting on the aircraft.

3.2.3 Drag on aircraft

The percentage change in skin friction drag is caused primarily by the increase in the kinematic viscosity of air, within the cruising space, which has a temperature lapse rate of about $8 \times 10^{-8}\ m^2\ s^{-1}\ K^{-1}$. **Figure 8c** indicates that, for all operating airliners considered, there could be a 3.5% increase in skin frictional drag by 2100 ($\Delta\tau/\tau^0 = 0.035$, with superscript means the value at reference year 2010), whereas the skin friction drag is only ~5.7% of the total drag. The increase in skin frictional drag accounts only for a ~0.2% reduction in efficiency in fuel consumption. Thus, due to increased air viscosity and decreased engine overall efficiency, the annual fuel consumption in 2100 would be ~0.9% higher than around 2010. The spread in the estimation is wide among climate models, but all indicate more fuel consumption as climate warms. The corresponding absolute change of 0.9% reduction in efficiency in fuel consumption is considerable, about 0.68 billion gallons of fuel annually. The reduction in thermal efficiency is complementary to the IPCC AR5 perspective, but the fact that the increased drag and mechanical efficiency may be a supplant concept (a new rubric) will hopefully stimulate further studies in this

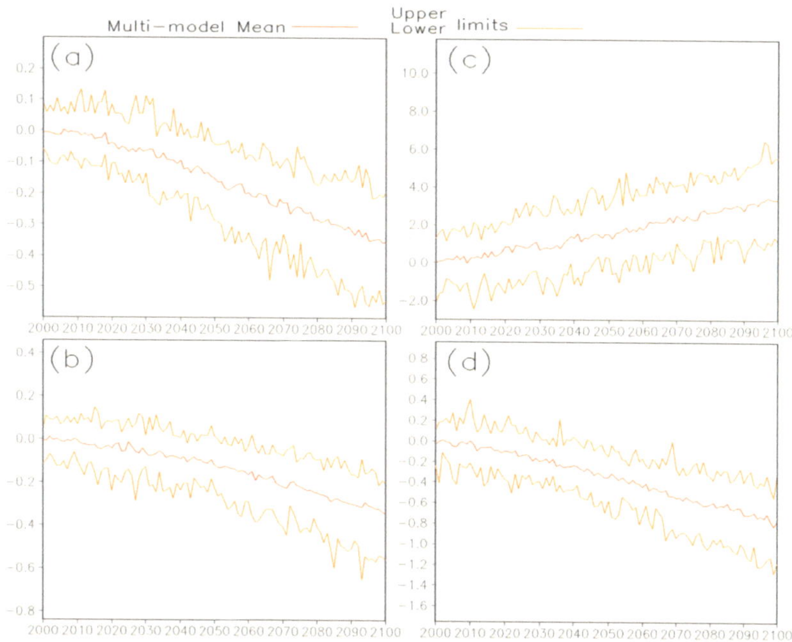

Figure 8.
Decreases in thermal efficiency (a) and mechanical efficiency (b), increase in skin frictional drag (c), and the overall decrease in fuel efficiency (d) during 2000–2100, for entire commercial aviation sector as a whole. Multiple climate model ensemble means (shown as thick red lines), and the ranges of the variability (thick yellow lines) are shown for 24 climate models under RCP 8.5 emission scenarios (for clarity only, the other two models both are within the range). The flight schedules of year 2010 are assumed unchanged during the entire period. Note that the control period is centered on 2010 (2005–2015), so the values at starting (2000) is not exactly united.

direction. A *t*-test was performed for the overall decrease in fuel efficiency time series. For 22-year periods centered at year 2010 and 2090, a *P*-value of 0.0017 (*dof* = 38) was obtained. At this significant level, it means at a possibility of 99.84%, the trend is not by mere coincidence. Thus, the net decrease in fuel efficiency is small but statistically significant.

3.3 Conclusion

To conclude, factors affecting aviation fuel efficiency are thermal and propulsive efficiencies and overall drag on aircrafts. An along-the-route integration is made for all direct flights in baseline year 2010, under current and future atmospheric conditions from nine climate models under the representative concentration pathway (RCP) 8.5 scenario. Thermal and propulsive efficiencies are affected oppositely by environmental warming. The former decreases 0.38%, but the latter increases 0.35% over the twenty-first century. Consequently, the overall engine efficiency decreases only by 0.02%. Over the same period, skin frictional drag increases ~5.5%, from the increased air stickiness. This component is only 5.7% of the total drag, the ~5.5% increase in air viscosity accounts for a 0.275% inefficiency in fuel consumption, one order of magnitude larger than that caused by engine efficiency reduction. The total decrease in fuel efficiency equals to ~0.24 billion gallons of extra fuel annually, a qualitatively robust conclusion but quantitatively with significantly inter-climate model spread.

The effects on fuel cost from increased airplane potential energy still is one order of magnitude smaller than factors considered here, due to the fact that it is a less than a 1% increase in the climbing stage (at most 1 hour). The fuel cost, in the cruising stage is much greater, because of the length of time (up to 14 hours). Also, at higher altitudes, the fuel cost for climbing is reduced (the increased tropopause height's effect is the upper level part of the trajectory). The common statement that the climb stage is more fuel-consuming refers to the rate, not the total value (except for very short flights, e.g., from Oklahoma City to Tulsa).

Climate effects on aviation are a burgeoning but promising research field. Our study here focused on the rudimentary aspects that are of concern to the commercial airlines: the effects on maximum payload and on fuel costs. Other directions such as customer comfort and safety also are profoundly affected, especially the circulation changes (winds and turbulence [24]). These will be addressed in future studies in this walk of line.

Acknowledgements

We are thankful for the useful discussions with Professors Huiling Yuan (Nanjing University), Weidong Guo (Nanjing University), and Zhaohua Wu (FSU) on the relevance of this research to environmental change and societal adaptations. We also thank Professors Mervyn Lynch (Curtin University), Zhaomin Wang (Hihai University), and Xiangbai Wu for providing constructive comments on the aspects of the manuscript.

Conflict of interest

The authors claim no conflict of interest in this research.

Author details

Diandong Ren[1] and Lance M. Leslie[2]*

1 Curtin University, Perth, Australia

2 University Technology, Sydney, Australia

*Address all correspondence to: lmleslie@ou.edu

References

[1] Minnis P et al. Global distribution of contrail radiative forcing. Geophysical Research Letters. 1999;**26**:1853-1856

[2] IPCC. Understanding the Climate System and its Recent Changes. In: Summary for Policymakers. IPCC AR5 WG1; 2013

[3] IPCC. Aviation and the Global Atmosphere—A Summary for Policy Makers. UK: Cambridge University Press; 1999. p. 373

[4] Sausen R et al. Aviation radiative forcing in 2000: An update on IPCC (1999). Meteorologische Zeitschrift. 2005;**14**:555-561

[5] Fichter C, Marquart S, Sausen R, Lee D. The impact of cruise altitude on contrails and related radiative forcing. Meteorologische Zeitschrift. 2005;**14**: 563-572

[6] Fu Q, Liou KJ. Parameterization of the radiative properties of cirrus clouds. Atmospheric Sciences. 1993;**50**: 2008-2025

[7] Stuber N, Forster P, Radel G, Shine K. The importance of the diurnal and annual cycle of air traffic for contrail radiative forcing. Nature. 2006;**441**: 864-867

[8] Lewis S, Karoly D. Are estimates of anthropogenic and natural influences on Australia's extreme 2010–2012 rainfall model-dependent? Climate Dynamics. 2015;**45**:679-695

[9] Holman J. Thermodynamics. New York: McGraw-Hill; 1980. p. 217

[10] Wallace J, Hobbs P. Atmospheric Science. 2nd ed. Burlington, MA, USA: Academic Press; 2006. p. 504

[11] Taylor K, Stouffer R, Meehl G. An overview of CMIP5 and the experiment design. Bulletin of the American Meteorological Society. 2012;**93**:485-498

[12] Kistler R et al. The NCEP–NCAR 50-year reanalysis: Monthly means CD-ROM and documentation. Bulletin of the American Meteorological Society. 2001;**82**:247-267

[13] Lawrence M. The relationship between relative humidity and the dew point temperature in moist air: A simple conversion and applications. Bulletin of the American Meteorological Society. 2005;**86**:225-233

[14] Trenberth K. Changes in precipitation with climate change. Climate Research. 2011;**47**:123-138

[15] Allen M, Ingram W. Constraints on future changes in climate and the hydrologic cycle. Nature. 2002;**419**: 224-232

[16] Held I, Soden B. Robust response of the hydrological cycle to global warming. Journal of Climate. 2006;**19**: 5686-5699

[17] Khon V et al. Response of the hydrological cycle to orbital and greenhouse gas forcing. Geophysical Research Letters. 2010;**37**:L19705

[18] Coffel E, Horton R. Climate change and the impact of extreme temperatures on aviation. Weather, Climate, and Society. 2015;**7**:94-102

[19] Hane F. Comment on "climate change and the impact on extreme temperatures on aviation". Weather, Climate, and Society. 2016;**8**:205-206

[20] Santer B et al. Behavior of tropopause height and atmospheric temperature in models, reanalyses, and observations: Decadal changes. Journal of Geophysical Research. 2003;**108**(D1): 4002. DOI: 10.1029/2002JD002258

[21] Reichler T, Martin D, Sausen R. Determining the tropopause height from gridded data. Geophysical Research Letters. 2003;**30**:2042. DOI: 10.1029/2003GL018240

[22] Kershner W. The Advanced Pilot's Flight Manual, 392. 8th ed. Newcastle, WA, USA: Aviation Supplies and Academics, Inc; 2015

[23] Cumpsty NA. Jet Propulsion. Cambridge, United Kingdom and New York, NY, USA: Cambridge University Press; 1997

[24] Ren D. Effects of global warming on wind energy availability. Journal of Renewable and Sustainable Energy. 2010;**2**:052301. DOI: 10.1063/1.3486072

Chapter 10

Red Wings Proposed by Robert Bartini for Sustainable Aviation

Sergej Težak

Abstract

This chapter is a brief description of aircraft designer Robert Bartini and his role in the development of the military, passenger, and transport aviation. Robert Bartini was educated in Austria-Hungary and Italy, and graduated from Milan Polytechnic Institute. In 1923, he fled Italy to escape Fascism and emigrated in the Soviet Union. After the First World War, the young Soviet Union (USSR) desperately needed new engineers and scientists who would provide the new country with development of modern industry and transportation. At that time, Western European countries had knowledge and experiences, especially in the field of aviation. In addition, the chapter presents Bartini's vision of the sustainable intercontinental and continental high-speed transport, which was the focus in the last years of his work and creation. The term "red wings" in the title of the article is related to his work in the "red" Soviet Union. In Russia, Robert Bartini is very popular as a researcher and developer. There are many books about him in Russian and Italian language, but not in English. Thus, his work is still quite unknown in the West. He was born in Kanjiza (today Serbia) in 1897 and spent his youth in Fiume (Rijeka, today Croatia).

Keywords: Bartini, air transport, aviation, development of aircraft, ekranoplan, ground effect, WIG crafts, Rijeka, Soviet Union

1. Introduction

During the cold war, the superpowers, the United States (USA) and the Soviet Union (USSR), wanted to demonstrate their achievements in the development of military technique and technology. Furthermore, both superpowers wanted to show their best in the field of engineering and technology for military and civil purposes. This competition reflected the conquest of space and the field of aviation, where the former Soviet Union wanted to be presented abroad in the best light. However, the knowledge in this area could not be obtained in a short time, so it was necessary to invest years and years into the development and testing and learning from mistakes. Due to the lack of prior knowledge, both superpowers were hungry for scientists and researchers, mainly from Europe, that contributed an important part in the development of space technology and aviation. Most of the necessary knowledge and human resource superpowers were gained after the Second World War. However, many scientists fled from the unstable pre-war Europe to the United States before the war, rather than the Soviet Union.

After the First World War, the Soviet Union wanted quicker progress in the field of aviation because of the country's vast size. The most famous pioneers in space transportation (Tsiolkovsky) and builders of large airplanes (Sikorsky, Tupolev) came from Russia, but knowledge in Western Europe in the field of aerodynamics made great progress, especially Italy which at that time possessed a superior knowledge in aerodynamics, which was rewarded with a world speed record in aviation in the years 1927–1929 and 1933–1934. Therefore, it is not surprising that Robert Bartini immigrated to Russia in 1923. He was a young Italian aviation engineer and communist proving his skills in the fields of aviation in the next 50 years in the Soviet Union. This case shows a classic transfer of knowledge in the new young Soviet Republic.

His solutions, gained with better knowledge of aerodynamics, have contributed to higher aircraft speed and lower energy or fuel consumption, and at the same time, a longer range of aircrafts was reached. In the historical context, his work has led to the improvement in the sustainability of aviation.

2. Life and work of Robert Bartini in the field of aviation

Robert Bartini was born on May 14, 1897, in Kanjiza (today this is the town in Serbia near Hungarian border), according to his documents from the time he lived in the Soviet Union [1]. When he was 3 years old, he was adopted by the family of rich state official in the town of Fiume in the Austro-Hungarian monarchy (today Rijeka in Croatia). In his youth, his surname was Orosdy [2]. As a young boy, he was very intelligent, and he received additional education by the family teacher in natural science (chemistry, biology), music, and foreign languages [3]. In 1912, he saw an airplane and was fascinated by the air show of Russian aviator Slavorossov in Rijeka. In 1915, he graduated from grammar school in Budapest [2], was drafted and sent to reserve officer school, located in the town of Bystritsa (in Czechoslovakia), and then in 1916, he was sent to Russian-Austrian-Hungarian front where he was captured in June 1916 [4]. He was sent into captivity in the Far East in prison camps in Khabarovsk and Vladivostok. He was released after the First World War, and then he worked as a taxi driver in Shanghai. In 1920, he returned home in Fiume. When working at Isotta-Fraschini facilities, he graduated from Milan Polytechnic Institute in 1922. In the same year, he also graduated from pilot school in Rome. During the captivity in Russia, Bartini came under the influence of communist literature, and in 1921, he became a member of the Communist Party of Italy. In 1923, the Fascists took over the control in Italy, and Bartini was sent to the Soviet Union. When he left Italy over Alps, he vowed, "red planes will always be faster than black" [3]. When he was a member of the Communist Party of Italy, he worked underground and got the nickname "Red Baron," which he kept from the Soviet Union as well.

Since the arrival in the Soviet Union, Bartini served the next 6 years as a mechanical engineer and the head of the department of scientific and test aerodrome (now Chkalovsky). Later, he was transferred to the First Squadron of Naval Forces of the Black and Azov Seas. In 1925, he participated in the national gliding championship in Crimea and together with designer Myasitchev cooperated as a constructor and pilot [1]. In 1928, he became the head of the department of amphibious experimental aircraft design and was appointed as the chief engineer of the Black Sea Aviation. In the following year, he participated in the organization of ANT-4 "Soviet Country" aircraft flight to the United States, servicing in sea segment of the route. In the Red Army, Bartini reached the rank of a brigadier (Brigadier General).

Bartini returned to Moscow and worked as a chief designer of his projects in seaplane design bureaus at facilities No. 22 and 39 until August 1930. He was dismissed from the design bureau due to his critical letter to Stalin, but the Air Forces authorities organized a small design bureau for him at facility No. 22. In this office, he began to design a new aircraft, Stal-6 (Steel-6) with an incredible clean contour (**Figure 1**). "I saw an airplane like a beautiful naked girl," said the former test pilot a half century later, when he saw the "top secret" aircraft for the first time [2]. The "Stal-6" aircraft was a monoplane configuration fighter airplane with inventive solutions: full retraction of bicycle landing wheel, a closed cockpit with a transmission gear in elevator control linkage, and an engine evaporative cooling system with cooler-wing tips. The structure and aerodynamics of this aircraft permitted the speed of 420 km/h in 1933 [5] which was 150 km/h more than the best fighters in the Soviet Union at the time. The aircraft had V-12 engine with 630 hp., and it was made of stainless steel "enerzh 6." Two years later (1935), Bartini designed the "Steel-8" fighter reaching the speed of 630 km/h, but it was not selected for the proposed Soviet fighter. If the Soviet Union had chosen this plane for further development, then they would have had a superior fighter for defending "mother Russia" from the Germans by 1941.

In 1934, Bartini began developing the Stal-7 aircraft, which was a twin-engine passenger aircraft, exhibited at the Paris Salon in 1936. In 1939, the aircraft reached a new world record for a distance of 5000 km; it was flying over 5068 km with an average speed 405 km/h. The top speed of this aircraft was 450 km/h. In fact, Bartini's knowledge of aerodynamics was most evident in the development of this airplane. Its special shape increased the pressure under a fuselage, thus enabling additional lift. In this way, the airplane consumed less fuel at higher speeds, which at those times contributed to sustainable aviation. The Stal-7 was ready for the flight around the world, which was prevented by the arrest of the chief designer Bartini. They imprisoned him in 1938 and accused of being Mussolini's agent and participating in the burning of building No. 240, where the aircraft Stal-7 was placed. First, he was sentenced to death and imprisoned in the disreputable NKVD prison. When his plane reached a world record, Stalin "personally" took care that Bartini's sentence was reduced to standard 10 years in prison. Bartini was transferred to secret research and development camp KB (sharaskas) in different towns: Moscow, Omsk, Kazan, and Taganrog. At that time, the Stal-7 aircraft was one of the greatest secrets of the Soviet Union, leaving behind only one short film clip and some photos. Under the leadership of V. Ermolajev and advice of Bartini, the plane began to transform into the long-range Yer-2 bomber, resulting in about 400 Yer-2 aircrafts.

During the Molotov-Ribbentrop pact, the aircraft was intended for attacks on Britain and France and their bases in the Middle East. Instead, the Yer-2 aircrafts

Figure 1.
Aircraft Bartini Stal-6 (Steel 6). Source: http://en.valka.cz/topic/view/102660/Bartini-Stal-6, web source: October 18, 2017.

were engaged in night-time attacks on Berlin at the time when the Germans were sure they were invincible. The first bombing took place already on August 8, 1941. In fact, the Soviets bombed Berlin directly from Moscow. At the beginning, the Germans did not even know what they were dealing with. Their fighters were simply too slow in 1941 (**Figure 2**).

During the period from 1940 to 1943, Bartini's main concern in the custody in Moscow and Omsk was faster-than-sound aircraft with rocket power. The project of his first P-114 (Cyrillic P = R for rocket) interceptor with a swept wing was not realized [4]. The P-114 was designed for speed more than 2000 km/h. At that time (1943), Bartini already knew that the best wing shape for speeds beyond Mach 2 is delta wings.

Then he started constructing the first wide-fuselage (wide-body) transport aircraft T-117 for transport tanks. This was the first aircraft with transport ramp/door at the back of the plane, for easier loading/unloading of cargo (**Figure 3**). The plane was already constructed in Taganrog, but necessary engines were not supplied. Those engines were required for the production of Tupolev Tu-4 bombers, which were a copy of the American B-29 bombers. Stalin said that Russia needed bombers

Figure 2.
Passenger aircraft Bartini Stal-7 and bomber Yer-2 (Er-2). Source: http://mig3.sovietwarplanes.com/colors/1945-50-oldtypes/yer2-splinter.jpg, web source: March 16, 2017.

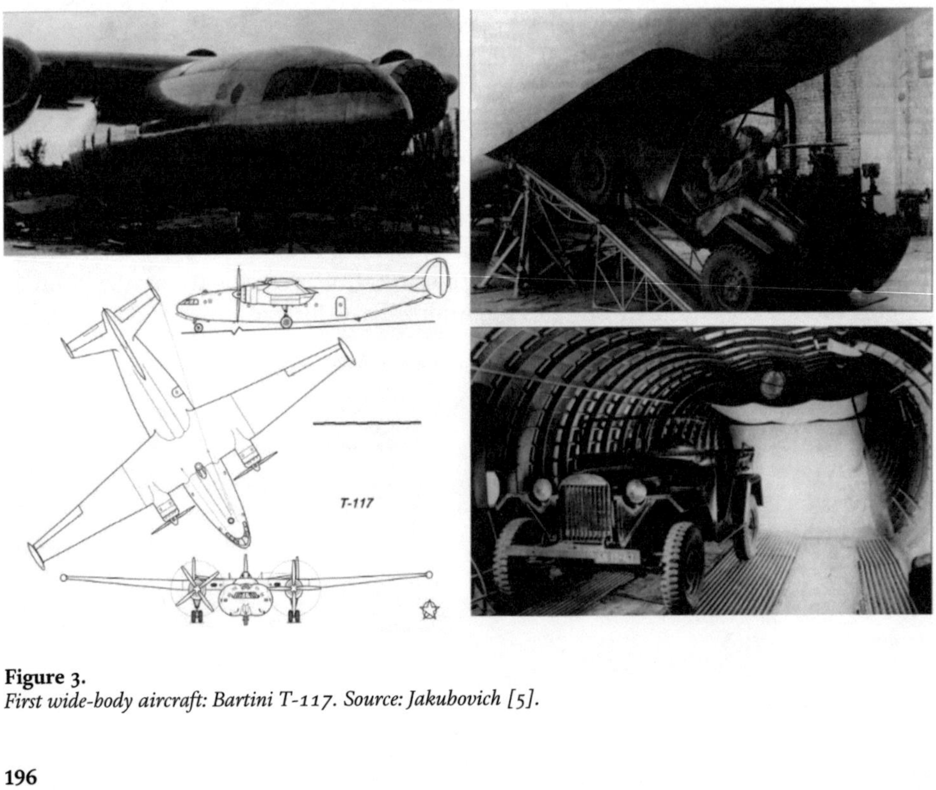

Figure 3.
First wide-body aircraft: Bartini T-117. Source: Jakubovich [5].

instead of transport aircrafts. In addition, new invented aircraft T-117 did not flight, so the project was laid off. T-117 aircraft drawings were sent to Antonov's company in Kiev. Many of Bartini's designs were later used on Antonov aircrafts. In 1946, Bartini was released, and his design bureau in Taganrog was closed; the T-117 plane was destroyed and cut. Antonov design bureau was the first wide-body transport aircraft constructed a decade later.

After the release, Bartini decided to continue the work in Taganrog, and he designed even more T-200 and T-210 transport aircrafts, but the realization of the projects failed to occur. In 1952, Bartini moved to Novosibirsk where he began to research and made aerodynamic calculation for the optimum shape of wings for supersonic speeds. Based on these investigations, he designed T-203 project—variable sweep wing with aerodynamic twist. For complex calculations, he used a computer BESM-1, the first Russian computer that was developed by S.A. Lebedev. Therefore, he began to develop the plans for strategic bombers A-57 and A-55 at speeds of 2200–2500 km/h with the possibility of landing on the water. At that time, Soviet bombers did not have sufficient range to reach the coast of the United States and return to the Soviet Union. Bartini found an innovative solution for aircraft landing on the water, where Soviet submarines could supply fuel to the aircraft. However, Soviet authorities refused the project in favor of the development of ballistic missiles. The results of Bartini's research and results about wings for large aircrafts in supersonic speeds were sent to the Tupolev design bureau. These solutions were used there to design the supersonic Tu-144 passenger aircraft, and the Concorde aircraft had the same wing design, too.

3. Bartini's vision of the sustainable intercontinental high-speed transport

Bartini was completely rehabilitated in 1957, and he returned to Moscow, where he worked in a small Kamov construction bureau. During this time, he began to study and compare the various forms of the transport and determine the most energy corresponding type to a certain speed.

Bartini developed a "Theory of intercontinental transport on Earth," which was completed in the 1960s. This sustainable theory takes into account the entire planet Earth for the implementation of transport services for ships, planes, helicopters, railways, etc. In addition, the interdependencies between the various criteria are considered: the amount of load, speed of delivery, the weather conditions, and the area required for various operations (stopping and moving vehicles, facilities for loading and unloading, etc.). He came to the solution that the most optimal and sustainable vehicle can fly just above the surface; it can take off and land vertically and can be applied on all surfaces—snow, water, earth, ice, and sand [6].

Similar research about the efficiency of the different means of the transport was launched by Von Karman in the 1950s of the last century. He noted that the "hydroglider" had the highest efficiency; in those times, this was probably the original term for the WIG vehicle or ekranoplan (Russian term). The term WIG vehicle or ekranoplan is used for vehicles that use ground effect—wing-in-ground effect (WIG effect) (**Figures 4** and 5).

The optimum flight is just above the flat surface where vehicles can take advantage of ground effect. Vehicles using ground effect achieve up to 30% more lift than normal planes at the same wing surface. Therefore, the ground effect enables less surface of the wings for the same lift force. Moreover, it has less drag, which is best seen in the diagram. Vehicles using WIG effect are sustainable because they have lower energy or fuel consumption, and therefore they can reach longer range (**Figure 6**).

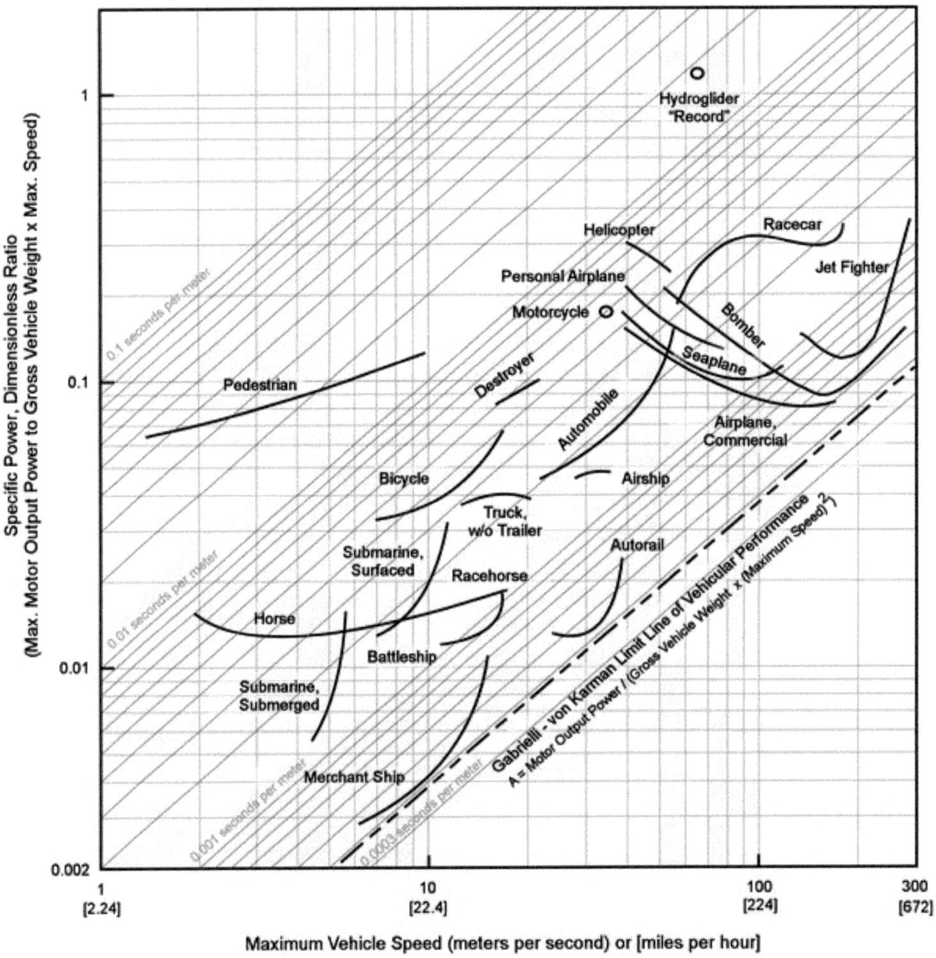

Figure 4.
Transport efficiency diagram for different means of transport. Source: Trains and boats and planes, https://trainsnboatsnplanes.wordpress.com/2010/01/07/the-price-of-speed/, web source: March 16, 2017.

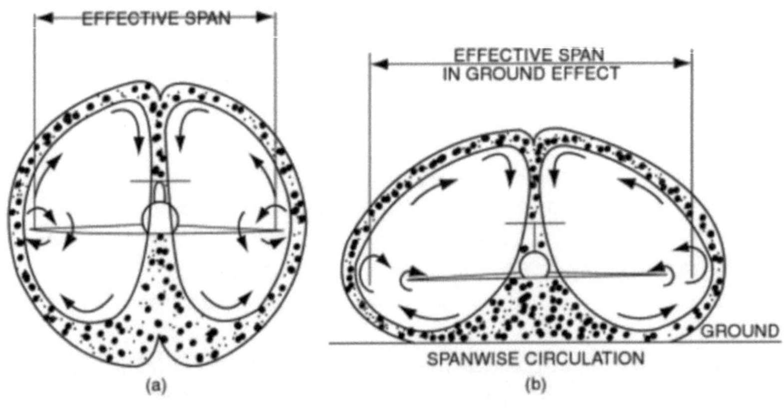

Figure 5.
Airplane in normal flight (a) and in fight with WIG effect (b). Source: Lun et al. [7].

In the 1960s, Bartini began to develop the VVA-14 prototype aircraft, an amphibian plane in order to prove his theory. This plane could land on all possible surfaces: sea, earth, ice, and sand. It had an option for vertical takeoff and landing or conventional takeoff and landing from both airports and water surfaces.

More importantly, WIG flight could save enough fuel for the vertical takeoff or landing of the aircraft. The plane was made in the Beriev factory. The first test flight was carried out in 1972, but 14 engines needed for vertical takeoff were never delivered. The plane made a series of test flights, took off on land and water in the conventional way (without testing vertical takeoff and landing). However, further development was not approved.

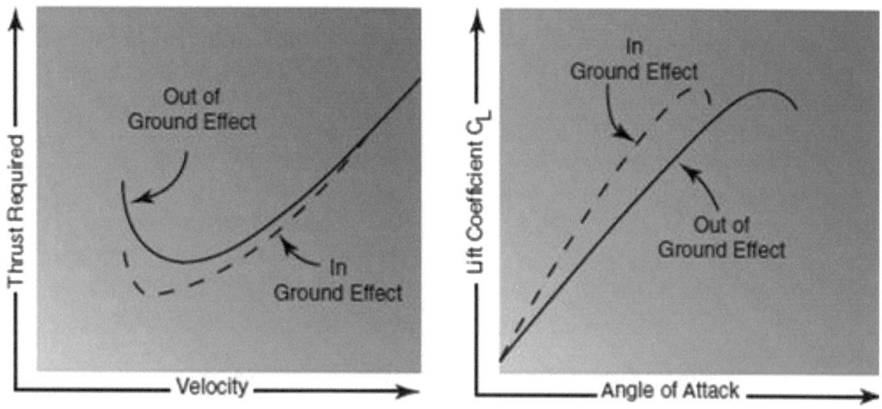

Figure 6.
Wing-in-ground effect flight changes drag and lift. Source: Aerodynamics in flight—Airplane ground schools, http://airplanegroundschools.com/Flight-Aerodynamics/index.html, web source: December 1, 2017.

Figure 7.
Amphibian airplane Bartini-Beriev VVA-14. Source: Jakubovich [5].

Figure 8.
Model of the WIG aircraft carrier and ekranoplan T-500. Source: Slavin [8].

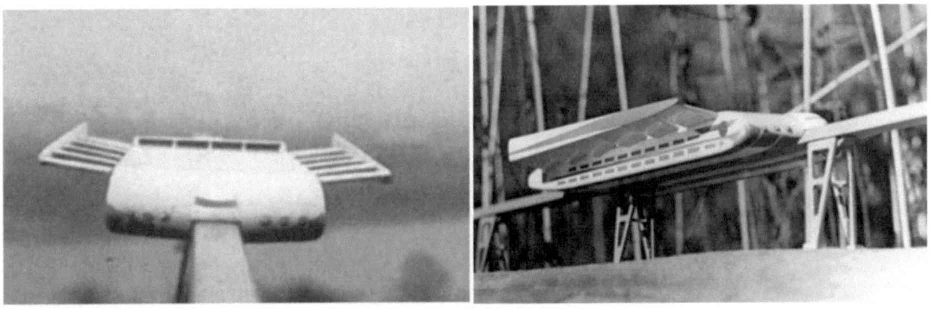

Figure 9.
Model of the future transport system (magnitoplan or ekranohod). Source: On Accasion of 110th Jubilee of Robert Bartini [4].

Bartini imagined a vehicle with greater utilization of ground effect, which could be used at high speed in transcontinental freight and passenger transport. Such ekranoplans would be more efficient than today's airplanes; they could transport more passengers and cargo and could actually still work in "friendlier" environment such as height above 10 km (**Figure** 7).

Bartini's design of WIG vehicles is probably the most efficient because the catamaran design with an open space between the floats, which accumulates compressed air, provides additional lift.

He continued with his ideas and designed an ekranoplan with weight of 2500 and 5000 tons, serving as an aircraft carrier and operating at speeds of 500 km/h. In such high speeds, aircrafts do not need a long runway for takeoff from the aircraft carriers. Aircraft carrier would travel at the same speed as the airplane (**Figure 8**).

Bartini was also considering quick continental transport. In the 1960s, he imagined a monorail vehicle traveling at high speed. Its performance is based on the compressed air such as hovercraft (ekranohod) [3] or on magnetic levitation (magnitoplan) [1]. The project was presented to the minister of transport B.P. Beschev and was also approved, but never realized. Bartini constructed additional aerodynamic surfaces at the sides of the vehicle to increase the lift or to control the correct distance of vehicle from the track (**Figure 9**).

4. Bartini's work in other fields of creation

In today's Russia, Robert Ludvigovich Bartini is described as a misunderstood genius whose ideas were ahead of his time, comparing him with Nikola Tesla. He is

DOI: http://dx.doi.org/10.5772/intechopen.85032

portrayed as the aircraft designer, physicist, astronomer, philosopher, painter, musician, and polyglot (he spoke seven languages and read nine).

He cooperated many times in his life with Sergei Korolev. When Sergei Korolev was appointed as the head of the Russian space program, he requested Robert Bartini as his mentor twice. Before this, Korolev often called Bartini his teacher. Firstly, Bartini was a head of Korolev for the first time before the Second World War, when he worked in Bartini's design bureau. Secondly, they cooperated in captivity (sharaska) when Bartini developed the rocket interceptor P-114. Sergei Korolev personally took care at the Soviet authorities about Bartini's project of the supersonic strategic bomber A-57. Korolev and Bartini also worked together on solutions to increase the range of the Myasishchev M-4 strategic bomber. Neither the presence nor the role of Bartini in advising the Soviet space program is yet fully known. In the last years of his life, he was primarily engaged in the exploration in physics, cosmology, and philosophy.

Bartini always tried to encourage innovative solutions. He gave an interesting answer to the question about what to do if the class filled with young professionals is given a problem, which should be solved in an innovative way. "Class must be extended," he responded. This means that many experts with their ideas from different fields of activity are more capable to get a better solution. Bartini was considering formulating mathematical method or model to determine the success of the idea or the patent. He developed a method AND-AND, which was based on recognized search solutions, which had already been used and therefore could predict the success of a given patent/idea for a new problem. He developed the method already in the 1930s, and it was called a method for the detection of talents. A similar but more general method was developed over 20 years later by Genrich Altshuller, who became famous with the TRIZ method. TRIZ method was spread to Western countries with the disintegration of the Soviet Union and is now quite well known.

The whole time of working in the Soviet Union, Bartini was solving problems in aerodynamics in a special way. He said that mathematics was the most exact science, and there was no doubt in it, whereas physics is derived science and the physical findings change over the centuries. He most doubted about physical constants, which he considered to be dependent of the specific "time and space." In 1965, he published an article on this subject, entitled "The relation between physical constants" [9]. The English version of the article was published in 2005 [10]. He claimed that the universe takes the form of six-dimensional torus—three dimensions of space and three dimensions of time. Dimensions of time include the first dimension as a length of time (the duration of the existence of the object), the second dimension of time as the width of time (number of cases/copies/images of body (parallel worlds)), and the third dimension as height of time (the speed of time is different in each of the worlds) (**Figure 10**).

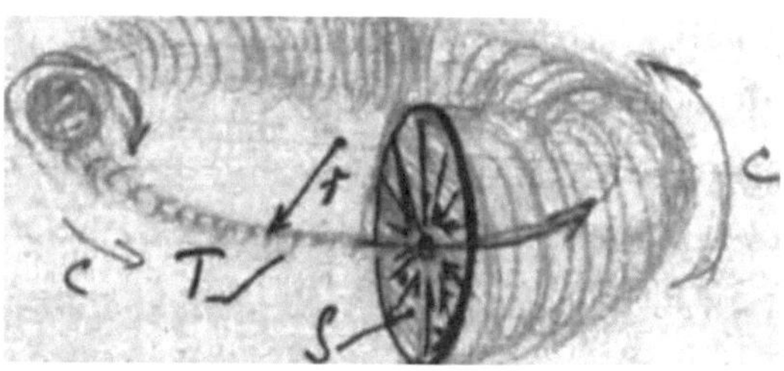

Figure 10.
Graphic presentation of the six-dimensional universe (Original Bartini's sketch). Source: Maslov [11].

The six-dimensional universe by Bartini is parallel streaming of times [1]. Of course, such revolutionary ideas caused him great problems, and they did not want to publish his article (he basically was not a physicist and was quite unknown in the world of science). However, Bruno Pontecorvo advocated for Robert Bartini and his article. Bartini gave to this article great importance; he signed as the author with the full name Robert Oros di Bartini [12]. It is interesting that the article even promised the proof of author's theory, but he soon died. Bartini used this physical theory to deal with his aerodynamic problems and came to excellent solutions. What is more, Bartini developed an entire philosophy on his physical theory of time. Therefore, his six-dimensional universe was named Bartini's universe. The second part of the article presents a table, a kind of periodic system of physical quantities expressed with the potentials of time and space, which Bartini developed together with Kuznetsov [13]. The most interesting part of this table is that some parts were an introduction into unpublished physical quantities such as the surface of time and the volume of time [14].

Bartini drew pictures in his lifetime; some were on the walls of his apartment for his well-being and better concentration at work. He had three children, two sons and a daughter and three grandchildren [15]. The first son Gero was a climber and died in 1959; the second son Vladimir lives in Taganrog and is an engineer. In 1967, Bartini got the Order of Lenin for his life achievements. In 1957, he got the Order of October Revolution.

Quite speculative and almost incredible or unbelievable but very interesting claims or mysterious stories appeared about the life and work of Robert Bartini in Russia [2]:

- Bartini in the 1920s and 1930s founded and headed a secret group ATON which members were renowned Soviet writers (sci-fi) and scientists.

- Bartini was as a prototype for Woland from the novel *The Master and Margarita*, since its writer Bulgakov was also a member of the ATON. The novel was written at a time when Stalin put in prisons and killed the "blossom" of Soviet intelligence (writers, engineers, scientists, officers), and during this time, Bartini was imprisoned as well.

- Bartini was also the prototype for the book *The Little Prince*. A writer Antoine de Saint-Exupéry and Robert Bartini met in their youth and both were impressed by aviation.

- Bartini was mentioned as the mysterious aircraft designer "Dunaev" that constructed the invisible plane.

- The whole story of Bartini's youth was invented, so the NKVD was very suspicious about him.

- Altshuller summarized the ideas of Bartini to develop his own method of TRIZ, as they were in contact.

- Robert Bartini and Leo Szilard—a physicist who discovered a chain reaction—were sitting at the same table in high school in Budapest.

- Before his death, Bartini wrote a will in which he demanded that his manuscripts are sealed and opened in 2197 (at 300th anniversary of his birth).

5. About the origin of the Robert Bartini

It is still quite unclear today where exactly was Robert Bartini from and who exactly his parents were. The official version from his biography in the book of Igor Chutko from 1978 [3] is that he spent his youth in Fiume and his father was vice-governor of Fiume Baron Lodovico Oros di Bartini, one of the richest and most distinguished people of the Austro-Hungarian Empire. His mother was a young barely 17-year-old girl, an orphan, who had an origin in a poor and impoverished Hungarian noble family from Miskolc. When she gave birth to her son, they told her that the baby died at birth, which was not true. In fact, Robert was secretly sent to a peasant family. The young woman did not believe and 6 months later she found her son. She wanted to show him to his father, the young Lodovico. However, when she came to Fiume with a child, she found out that her boyfriend had already married. She returned home at night, left her son and drowned at dawn. The child was again given to the same farmer family. Later, the farmer who adopted the child went to work as a head farmer at the residence of the vice-governor Lodovico Bartini, the child's father. His wife Paola could not have children, and she wanted to adopt the young 3-year-old Robert, but then the gardener suddenly escaped with a child. She was persistent, so she hired a detective and realized that the child's father was her husband. Later they adopted the child and raised him with all the love. Robert had his own teacher at home, family doctor Dr. Baltazar, who taught him natural science (chemistry, biology), music, and foreign languages.

However, researchers Olga and Sergei Buzinovsky discovered that this story has shortcomings. They wrote three books about Bartini after the fall of the communist system [1, 2]. They found that Bartini himself mentioned that his real mother's surname was Fersel (Fertsel), his real father was Austrian Baron Formaha, and he was born in Kanjiza. This information was obtained by the NKVD from Bartini, who was questioned in 1938 when he was imprisoned.

In their research, they got help with data from the Embassy of Croatia in Moscow and the workers of the city archives in Rijeka [2]. The director of the archive Goran Crnković reported that in September 1912, Russian pilot Chariton Slavorossov really flew in Fiume. It is possible that the famous pilot and young Robert Bartini actually met as stated in the biography [3]. However, the vice-governor until 1902 was Francesco Vio, who was then appointed as the Governor and vice-governor became Andrea Bellen. Information about people named Bartini, Formaha, and Fersel was not found in the archive, but they found another trail. Near Fiume there was the estate of Baron Phillippe (Fülöp) Orosdy, Italian born, large landowner, and deputy of the upper house of the Hungarian Parliament. The Baron was also in the list of honorary members of the Hungarian Aero Club. He had a brother Lajos (Italian, Ludovico; German, Ludwig) in Budapest. So who was the father of Robert? Is it Baron Phillippe or his brother Lajos? Probably Lajos, because Robert was bearing his name (Robert Ludvigovich Bartini). He spent his youth with his uncle in Rijeka and probably attended high school (gymnasium) in Budapest and lived there with his father. This is most likely so, because when Robert Bartini was in sharaska, in 1939, Karl Szilard recognized him there. Karl was a distant relative of the nuclear physicist Leo Szilard [16]. It turned out that as a child Robert Bartini was sitting at a desk in the classroom with Leo Szilard, famous physicist [2].

Another researcher is Giuseppe di Ciampaglia, who wrote a book *La vita e gli aerei di Roberto Bartini* [17]. In 2015 [14], he wrote that Robert Bartini was the adoptive (and perhaps natural) son of Lajos Orosdy, the captain of the Hungarian border police in Fiume, later the doctor and the advisor of the governor. He moved to Rome after the annexation to Italy. The restless but excellent high school student

Robert patented a "stabilizer" at the age of 18. On August 16, 1915, the governor of Fiume István Wickenburg recommended to the Honvéd (defense) minister to transferred Robert Bartini to aviation. He stayed in the infantry on the Russian front and finished in the prison in Siberia.

In October 2017, Croatian President Kolinda Grabar-Kitarović, during her visit to the Russian President Vladimir Putin, highlighted two important people from Croatia who helped developing Russia in the past. The first was Robert Bartini, aviation designer, scientist, and visionary; the other was Matija Zmajević, Admiral of the Baltic Fleet of the Emperor Peter the Great.

Who was Robert Bartini? Hungarian, Italian, Austrian, Russian, Croat, or Serb? He himself answered the question, "If every particle in the human body is replaced every few years, then I am surely the Russian, since I have been living in Russia for decades" [2].

6. Conclusion

The story of Robert Bartini is very interesting because he was unlike any other European scientists and researchers working in the Soviet Union. It is even more amazing that he was a son of powerful and rich Baron, and he decided for communism. Otherwise, the stormy first half of the twentieth century with two World Wars in Europe was not lenient to the fate of the people who are mutually intertwined in the most unusual combinations.

Early in his career, Bartini had luck and was rapidly promoted in the army and in the construction of aircraft, supported by his patron Marshal Tukhachevsky, former leader of the Soviet Armed Forces (Red Army). When Tukhachevsky, Bartini, and other officers were arrested before the Second World War, Bartini was lucky to survive. Since then, the authorities in Moscow did not support him any longer. After the war, he was able to realize only one idea—a prototype of the VVA-14 aircraft. With his work, Bartini was the most competitive of all his colleagues—Russian aircraft constructors. He was full of new ideas, so they were jealous, and probably Soviet authorities could not allow their planes to have an Italian name.

Most of his ideas contributed to better aerodynamics of aircrafts resulting in lower energy and fuel consumption. In this way, his airplanes could achieve higher speeds. His theory of intercontinental transport on Earth also took into account the smallest needs of transport infrastructure. His "red wings" significantly contributed to sustainable aviation.

As a designer, he managed to make fly only four aircraft prototypes; nevertheless, his ideas and solutions were used by other aircraft designers. Like many other geniuses, he was not suitable for the implementation of mass production of aircrafts, but he was always looking for something new. His work includes about 60 aircraft designs and ideas for about 200 aircrafts. As a genius, Bartini had similar characteristics to Nikola Tesla, namely, he formed his devices in his head and then just draw what he saw in his mind. Above all, he was acclaimed as a very good teacher, because he was happy to share his knowledge with younger Soviet designers, who later created very successful aircrafts such as Beriev, Korolev, Simonov (chief designer of Su-27), and others.

Author details

Sergej Težak
Faculty of Civil Engineering, Transportation Engineering and Architecture, University of Maribor, Maribor, Slovenia

*Address all correspondence to: sergej.tezak@um.si

References

[1] Buzinovsky S. RO (about the mysterious fate of Robert Bartini). Electronic Library. RoyalLib.ru [Internet]. 1993. Available from: http://royallib.ru/book/buzinovskiy_sergey/ro_o_zagadochnoy_sudbe_roberta_bartini.html [Accessed: 16 March 2017]

[2] Buzinovskaya O, Buzinovsky S. The Secret of Woland. Electronic Library. RoyalLib.ru [Internet]. 2003. Available from: http://royallib.ru/book/i_olga/tayna_volanda.html [Accessed: 16 March 2017]

[3] Chutko I. Red plane. Electronic Library. RoyalLib.ru [Internet]. 1982. Available from: http://royallib.ru/book/chutko_igor/krasnie_samoleti.html [Accessed: 16 March 2017]

[4] On occasion of 110th Jubilee of Robert Bartini. BERIEV Aircraft Company [Internet]. 2007. Available from: http://www.beriev.com/eng/Pr_rel_e/pr_110e.html [Accessed: 16 March 2017]

[5] Jakubovich NV. Aircrafts R.L. Bartini. Moscow: Rusavia; 2006

[6] Unnecessary plane—Amphibious aircraft vertical takeoff and landing VVA-14 (Part 1) [Internet]. 2013. Available from: http://topwar.ru/692-samolyot-amfibiya-vertikalnogo-vzlyota-i-posadki-vva-14.html [Accessed: 16 March 2017]

[7] Lun Y, Bliault A, Doo J. WIG Craft and Ekranoplan: Ground Effect Craft Technology. Available from: https://books.google.si/books?hl=sl&lr=&id=oAXEphDmmEMC&oi=fnd&pg=PR5&dq=ekranoplan&ots=T6nutRSzgJ&sig=AotV_9I-0bKwWhLt_0xIiXYK_-A&redir_esc=y#v=onepage&q=ekranoplan&f=false, web source: March 16, 2017

[8] Slavin A. The Bartini Code, The Riddle of the Red Baron, movie. 2005. Available from: https://www.youtube.com/watch?v=sncUEiWAnAA, web source: 17 March 2017

[9] di Bartini RO. Some relations between physical constants. Doklady Akademii Nauk USSR. 1965;**163**(4): 861-864

[10] di Bartini RO. Relations between physical constants. Progress in Physics. 2005;**3**:34-40 (in English) [Internet]. 2007. Available from: http://www.ptep-online.com/index_files/2005/PP-03-06.PDF [Accessed: 16 March 2017]

[11] Maslov AN. Universe Bartini, A Collection of Articles on Physics and Philosophy. Moscow: Samoobrazovanie; 2009

[12] Gerhstein SS. Recollections and Reflections about Bruno Pontecorvo [Internet]. Available from: http://pontecorvo.jinr.ru/gershtein.html [Accessed: 27 March 2017]

[13] Tennenbaum J. Pobisk Kuznetsov, a Modern Day Leonardo Da Vinci [Internet]. 2001. Available from: https://archive.schillerinstitute.com/russia/lar_hzl_121401/pobisk_leonardo.html#article [Accessed: 01 December 2018]

[14] Ciampaglia G. Krasnyj Baron [Internet]. 2015. Available from: http://www.societaitalianastoriamilitare.org/COLLANA%20SISM/Giuseppe%20Ciampaglia,%20Kransnyj%20Baron.%20Robert%20Ljudvigovich%20Bartini%201897-1974.pdf [Accessed: 01 December 2018]

[15] Information of the inaugural meeting of the 112th anniversary of birth of Robert Oros di Bartini held on

May 14th 2009 in the Memorial Museum of science Zhukovsky [Internet]. 2009. Available from: http://www.chronos.msu.ru/old/rnews/bartini_informatsiya.pdf [Accessed: 07 March 2017]

[16] Glinsky A. Theremin Ether Music and Espionage. Urbana and Chicago: University of Illinois Press; 2000

[17] Ciampaglia G. The Life and Planes of Roberto Oros di Bartini. 2nd ed. Rome: I.B.N. Publisher; 2010